Political Values and Narratives of Resistance

This book brings together multidisciplinary perspectives to explore how political values and acts of resistance impact the delivery of social justice in post-colonial states.

Everyday life in post-colonial states, such as South Africa and Zimbabwe, is characterized by injustices that have both a historical and contemporary nature. From fishers in Cape Town accused of poaching, to residents of Bulawayo demanding access to water, this book focuses on the relationship between the state and groups that have been historically oppressed due to being on the margins of the political, economic, and social system. It draws on empirical research from 12 scholars looking at cases in Brazil, India, South Africa, and Zimbabwe. Chapters explore questions such as what citizens, especially those from marginalized groups, want from the state. The book looks at the political values of citizens and how these are formed in the process of engaging with the state and through everyday injustices. It also asks why and how citizens resist the state, with examples of protest, as well as less visible forms of resistance reflecting complex histories and power relations. Finally, the book explores how narratives and counter-narratives reveal the nature of political values and perceptions of what is just. Taken together, these elements show the evolution of post-colonial social contracts.

Examining important themes in political science, anthropology, sociology and urban geography, this book will appeal to scholars and students interested in political values, justice, social movements and resistance.

Fiona Anciano is Associate Professor in the Department of Political Studies at the University of the Western Cape, South Africa. Her research areas include civil society, democracy, and urban politics.

Joanna Wheeler is a Marie Sklodowska Curie Fellow at the Centre for Trust, Peace, and Social Relations, Coventry University and Senior Research Fellow, Department of Political Studies, University of Western Cape. Her research interests include citizen participation and inclusion, gender and violence in urban contexts, and storytelling and participatory visual methodologies.

Routledge Research on Decoloniality and New Postcolonialisms

Series Editor
Mark Jackson, Senior Lecturer in Postcolonial Geographies, School of Geographical Sciences, University of Bristol, UK.

Routledge Research on Decoloniality and New Postcolonialisms is a forum for original, critical research into the histories, legacies, and life-worlds of modern colonialism, postcolonialism, and contemporary coloniality. It analyses efforts to decolonise dominant and damaging forms of thinking and practice, and identifies, from around the world, diverse perspectives that encourage living and flourishing differently. Once the purview of a postcolonial studies informed by the cultural turn's important focus on identity, language, text and representation, today's resurgent critiques of coloniality are also increasingly informed, across the humanities and social sciences, by a host of new influences and continuing insights for different futures: indigeneity, critical race theory, relational ecologies, critical semiotics, posthumanisms, ontology, affect, feminist standpoints, creative methodologies, post-development, critical pedagogies, intercultural activisms, place-based knowledges and much else. The series welcomes a range of contributions from socially engaged intellectuals, theoretical scholars, empirical analysts and critical practitioners whose work attends, and commits, to newly rigorous analyses of alternative proposals for understanding life and living well on our increasingly damaged earth.

This series is aimed at upper-level undergraduates, research students and academics, appealing to scholars from a range of academic fields including human geography, sociology, politics and broader interdisciplinary fields of social sciences, arts and humanities.

Making Urban Theory
Learning and Unlearning through Southern Cities
Mary Lawhon

Decolonising Schools in South Africa
The Impossible Dream?
Pam Christie

Housing in the Aftermath of the Fast Track Land Reform Programme in Zimbabwe
Lovemore Chipungu and Hangwelani Hope Magidimisha

Political Values and Narratives of Resistance
Social Justice and the Fractured Promises of Post-colonial States
Fiona Anciano and Joanna Wheeler

For more information about this series, please visit: https://www.routledge.com/Routledge-Research-on-Decoloniality-and-New-Postcolonialisms/book-series/RRNP

Political Values and Narratives of Resistance

Social Justice and the Fractured Promises of Post-colonial States

Edited by Fiona Anciano and Joanna Wheeler

Routledge
Taylor & Francis Group
LONDON AND NEW YORK

First published 2021
by Routledge
2 Park Square, Milton Park, Abingdon, Oxon OX14 4RN

and by Routledge
52 Vanderbilt Avenue, New York, NY 10017

Routledge is an imprint of the Taylor & Francis Group, an informa business

British Library Cataloguing-in-Publication Data
A catalogue record for this book is available from the British Library

Library of Congress Cataloging-in-Publication Data
A catalog record has been requested for this book

ISBN: 978-0-367-63903-7 (hbk)
ISBN: 978-0-367-63905-1 (pbk)
ISBN: 978-1-003-12124-4 (ebk)

Typeset in Times New Roman
by SPi Global, India

Contents

Figures

Acknowledgements

This volume reflects a long journey, in terms of the ideas and relationships that have evolved in the process of completing it. As editors, we would like to thank all the contributors, first and foremost, for their commitment and efforts. Through the many discussions, rounds of feedback, and editorial requirements, the contributors in this volume have persevered. This determination shows in the strength of the ideas that they have each communicated in their chapters. They have also generously given their time to provide feedback on other chapters in the volume as well as the introduction.

The idea for this volume first emerged at a Bellagio Center workshop in Italy, which was generously sponsored and hosted by the Rockefeller Foundation in 2017. We were then able to bring together the contributors to this volume at a second workshop in Cape Town, South Africa, in 2018, with the support of the South African National Institute for Humanities and Social Sciences. Each of these events contributed to the intellectual development of the ideas in this volume and facilitated the space for a collaborative approach. We are very fortunate to have had these opportunities as the foundation for this volume.

We are also grateful to the three peer reviewers who provided critical and constructive feedback on an earlier draft. A big thank you to our copy editor, Diemo Masuko. Her professionalism and skilled support has been invaluable to complete this volume.

Finally, we would like to thank our respective families for supporting us, and for not losing their patience with the many evenings, weekends, and early mornings when working on this book took over everything else.

Abbreviations

AGM	Annual General Meeting
ANC	African National Congress
BCC	Bulawayo City Council
BPRA	Bulawayo Progressive Residents Association
BRICS	Brazil, Russia, India, China, and South Africa
BSAC	British South African Company
BWC	Bulawayo Waterworks Company
CBLE	Community-Based Legal Entities
CIMI	Indigenist Missionary Council
CORD	Collaboration for Research on Democracy
COVID-19	Coronavirus Disease-2019
CPI	Communist Party of India
CRPF	Central Reserve Police Force
CSO	Civil Society Organisations
DA	Democratic Alliance
DAFF	National Department of Agriculture, Forestry and Fisheries
DSEI	Special Indigenous Health District
EFF	Economic Freedom Fighters
FRA	Forest Rights Act
G20	Group of Twenty
HBCA	Hout Bay Civic Association
HIV/AIDS	Human Immunodeficiency Virus/Acquired Immuno-Deficiency Syndrome
IBSA	India, Brazil and South Africa
ICM	Idealized Cognitive Model
IDC	International Development Cooperation
IFI	International Financial Institution
INDAL	Indian Alumina Company Limited
LGBTIQ+	Lesbian, Gay, Bisexual, Transgender, Intersex, and Queer
LPG	Liberalization, Privatization and Globalization
MDC	Movement for Democratic Change
NALCO	National Aluminum Company Limited
NDB	New Development Bank

NGO	Non-Governmental Organisation
NSS	Niyamagiri Suraksha Samiti
PESA	Panchayats Extension to Scheduled Areas Act
PF-ZAPU	Patriotic Front-Zimbabwe African People's Union
PIL	Public Interest Litigation
PPWM	Pre-Paid Water Meters
PSSP	Prakrutika Sampada Surakha Parishad
PT	Partido dos Trabalhadores (Worker's Party)
RTC	Reclaim the City
SACP	South African Communist Party
SSC	South-South Cooperation
SSF	Small-Scale Fishing
SUS	Sistema Único de Saúde
TAC	Total Allowable Catch
UAIL	Utkal Alumina International Limited
UNDP	United Nations Development Program
WHO	World Health Organisation
WSS	Women Against Sexual Violence and State Repression
ZANU-PF	Zimbabwe African National Union-Patriotic Front
ZINWA	Zimbabwe National Water Authority

Contributors

Fiona Anciano is Associate Professor in the Department of Political Studies at the University of the Western Cape, South Africa. Her research areas include civil society, democracy and urban politics. Her publications include *Democracy Disconnected: Participation and Governance in a City of the South* (co-authored with Laurence Piper, Routledge, 2019).

S.J. Cooper-Knock is Lecturer in International Development at the University of Edinburgh. She is jointly appointed between the Centre of African Studies and Social Anthropology. She researches policing and punishment, statehood and citizenship in South Africa. She is an associate researcher at the University of the Witwatersrand.

Mmeli Dube is Research Assistant in the Political Studies Department at the University of the Western Cape. His research areas include urban politics, civil society, and citizenship. He is an alumnus of the Hubert Humphrey Fellowship on Public Policy at the University of Minnesota, Twin Cities. He is currently a PhD candidate at the University of the Western Cape.

Ranjita Mohanty is an Indian sociologist based in New Delhi. She researches development, social movements, social policies, and governance. She has a PhD from Jawaharlal Nehru University, New Delhi. She was a Fulbright Fellow at the University of North Carolina, Chapel Hill. She has published extensively, including books.

Nobukhosi Ngwenya is a Junior Research Fellow at the African Centre for Cities, which is an interdisciplinary research and teaching programme based at the University of Cape Town. Nobukhosi's research focus is community-based planning and development of human settlements.

Yusra Price is an anthropologist, researcher, and facilitator. She holds a master's in anthropology from the University of Cape Town (UCT) and has research interests in education and storytelling. She designs and facilitates participatory workshops for students in the UCT Anthropology Department to build ethnographic research skills. Together with Joanna Wheeler, she is also a member of the Ibali network hosted through Open University that is currently focused on Storytelling for Transformation.

Vera Schattan P. Coelho is Senior Researcher and Team Coordinator at the Brazilian Center of Analysis and Planning (Cebrap) and an associate researcher at the Centre for Metropolitan Studies (CEM, University of São Paulo). She teaches at the Federal University of ABC (UFABC), São Paulo, Brazil. She works with both qualitative and quantitative research methods and has extensive experience in evaluating policies and coordinating research projects at the local, national, and international levels (Brazil, Mexico, Argentina, Uruguay, India, Bangladesh, South Africa, Mozambique, Nigeria, Kenya). She has led various comparative studies in the areas of pension reform, health systems, and citizen involvement in public policies. The results of these studies have been presented and published nationally and internationally.

Katharina Schramm holds the Chair for Social and Cultural Anthropology at the University of Bayreuth. Her research interests focus on global inequalities, political subjectivities, and the politics of difference, with a particular emphasis on race. She approaches these questions through a material-semiotic approach.

Shylashri Shankar is Senior Fellow at the Centre for Policy Research in New Delhi. Her books include *Scaling Justice: India's Supreme Court, Anti-Terror Laws and Social Rights*, *Battling Corruption: Has NREGA Reached India's Rural Poor?* (co-authored with Raghav Gaiha), *A Secular Age beyond the West: Religion, Law and the State in Asia, the Middle East and North Africa* (co-edited with Mirjam Kunkler and John Madeley), and *Turmeric Nation: A Passage through India's Tastes*. She has held academic positions at the University of Texas at Austin and the Center on Religion and Democracy, University of Virginia. She was a Bellagio Fellow, and a co-convener of an international research cluster at ZiF, University of Bielefeld.

Bettina von Lieres is Assistant Professor in the Centre for Critical Development Studies at the University of Toronto Scarborough, Canada. She teaches courses in the field of critical citizenship studies. From 1991 to 2002 she held university positions in South Africa. She is currently appointed as an Extra-ordinary Senior Researcher at the University of the Western Cape, South Africa.

Laura Trajber Waisbich is a political scientist and international relations expert working on citizenship and development in the Global South. Laura is currently affiliated with the Brazilian Centre for Analysis and Planning (Cebrap) and the South-South Research and Policy Centre (ASUL) in Brazil. Laura holds a research master's from the Paris Institute of Political Sciences (Sciences Po Paris). She is also a PhD candidate at the University of Cambridge, United Kingdom.

Joanna Wheeler is a Marie Sklodowska Curie Fellow at the Centre for Trust, Peace, and Social Relations, Coventry University, and is Senior Research Fellow, Department of Political Studies, University of Western Cape. Her research interests include citizen participation and inclusion, gender and violence in urban contexts, and storytelling and participatory visual methodologies. Recent publications include Wheeler, J. and Bivens, F. (2020) 'Storytelling as Participatory Action Research' in Burns, D., Howard, J. and Ospina, S. (eds.), *SAGE Handbook of Participatory Action Research*, Oxford: Sage.

Foreword

Why do people protest, and what do these acts of defiance and resistance mean for broader democratic relationships between citizens and authorities which affect their lives? Across ten compelling cases in four 'post-colonial' states (Brazil, India, South Africa and Zimbabwe), the essays in this volume give a powerful and nuanced understanding of the lived experience of marginalised groups, how their actions are shaped by their everyday sense of what is fair, and what this means for often lofty and abstract discussions of concepts like the 'social contract', 'justice', 'participation', or the 'moral economy.'

There are a number of reasons why this book is so compelling. One is simply the diversity of cases, which provide a rich tapestry of why and how often marginalised groups of citizens protest and resist across a range of issues, be they around water and land, health and housing, livelihoods and legal processes, even foreign policy and global development issues. By showing how these issues are understood by ordinary citizens, and how they affect their everyday lives, the cases then call attention to the multiple ways in which citizens may act, whether in visible protests or more hidden forms of resistance, through online 'spectator citizenship' or deeply hidden 'silent citizenship', challenging closed spaces or through claiming and re-claiming existing spaces, or telling their own stories or 'counter story telling' to bring alive and make visible subaltern voices.

Often in volumes such as these, such a diversity of cases – across contexts and issues, through multiple methods and disciplinary lenses – risks obscuring the common threads of the collection. But in this case, each chapter comes back, again and again, to several core themes, which are eloquently articulated and woven together in the introductory chapter by the editors. To me, three of these themes stand out in particular.

First, these cases go behind the acts of citizenship, whether though protest or resistance, to understand the political values which motivate and animate them, and how such values are cultivated, shaped and change over time. Through taking the lens of grounded, lived experience of marginalised groups, we come to understand the people behind the protests in new ways – no longer as simply 'angry' or 'aggrieved', 'poor' or 'marginalised', 'left out' or 'left behind', but as active and thoughtful agents who have deep values of fairness, a sense of what is just, and an expectation that the liberal developmental and political promises of their post-colonial democracies should be fulfilled.

Secondly, these values and expectations have emerged organically and historically, often built as part of the process of constructing states in a post-colonial period which also popularised a new vision of what post-colonial democracies could be – more inclusive, more developmental, more equitable, with an ability to meet basic human needs and rights. These visions offered the foundations of a new social contract, the pact between states and citizens that builds legitimacy, is held together by trust and which provides the basic understanding of rights and responsibilities of citizens and their leaders. But while the social contract is often understood in legalistic terms, the cases in this book help us understand that the social contract is embodied in a set of norms and expectations as well, often unwritten as well as written, tacit as well as expressed, and interpreted differently by differently situated groups over time and place. When the social contract becomes fractured, when the state and authorities no longer deliver, then these groups will raise their voices through protests and resistance in a myriad of ways.

Third, while the cases in this book focus on what is termed 'post-colonial' states, by no means are its insights limited to these settings. Rising global inequality, intersecting around race, wealth, and gender and made more legible by the COVID-19 pandemic, global austerity and the climate crisis – all of these have strained the social contract in states around the world, from so-called developed democracies to more newly emerging ones. We have learned that the social contract is fragile and easily fractured, across contexts and states. These fractures risk growing fissures amongst the citizenry as well, as people give up their trust or belief in the legitimacy in democratic institutions and processes, turn their grievances against one another though intolerance and violence, or become victims of populist manipulations and disinformation.

But these essays remind us that there is also another more positive story behind these fractures – of citizens who have deeply engrained values of fairness and justice, and who have the courage to speak out and resist for what they believe to hold institutions accountable for their end of the democratic bargain. Protests and resistance as we see them here do not represent the failures of democracy; rather, they offer the best hope that it can be renewed in the face of the challenges it faces.

For me, reading this volume has been an inspiration in another way as well. Many of the contributors to this volume, including one of the editors, were originally part of another global network known as the Development Research Centre for Citizenship, Participation and Accountability, which I had the privilege of directing from 2000 to 2010. When the funding for that programme ended, a number of the southern-based researchers in the network had the determination to keep the community going, through a new network called the Collaboration for Research on Democracy (CORD) which continued to meet around the world, often with little or no funding, and co-constructing a research agenda relevant to the needs and challenges of their own societies. A decade later, this volume is one of the products of this network of researchers, whose deep dive into political values, protest, and resistance in these post-colonial countries gives insights for those of us interested in understanding the power of citizen action in whatever part of the world we might find ourselves. Their commitment to engaged and collaborative scholarship in order to understand and illuminate the challenges of contemporary democratic citizenship demonstrates their own political values and beliefs in the possibility of a fairer and more just world.

John Gaventa, Professor
Institute of Development Studies
August 2020

Introduction

Seeking social justice in the post-colonial state

Fiona Anciano and Joanna Wheeler

Introduction

Democratic political settlements in Brazil, India, South Africa and Zimbabwe led to constitutional promises of social justice, development, equality and inclusion. Mounting citizen discontent over the past decade, however, shows that the promises of these new social contracts now ring hollow for many. From fishers in Cape Town, South Africa, accused of poaching; to residents of Bulawayo, Zimbabwe, demanding access to water; everyday life is characterised by injustices of a historical and contemporary nature. This volume draws on empirically rich chapters from 12 authors looking at cases across Brazil, India, South Africa and Zimbabwe. These cases are in the 'global South'[1] and are in moments of significant political and economic change, where citizens[2] are rebelling against perceived failures of the state to provide services and access to livelihoods based on constitutional promises.

These moments of contestation are the subject of extensive research and analysis in terms of how they can be understood as participation or resistance, as well as their consequences for democracy (Gaventa and McGee 2010; Thompson and Tapscott 2010). Moving beyond analyses of social movements and deepening democracy, this volume brings fresh insights that suggest that there is a more fundamental challenge to the social contract itself. The volume does this by looking at the relationship between political values, resistance and the social contract. It shows how fractures in the social contract deepen where there are perceptions that social justice has not been delivered. The volume looks at case studies of individuals and groups who live with everyday forms of injustice, and thus our analysis focuses on the relationship between the state and groups that have been historically oppressed because they are often on the margins of the political, economic and social systems. Through a focus that spans the individual and systemic levels, we consider a series of interrelated questions, prompted by events unfolding in each context.

First, we ask: What do citizens, especially those from historically oppressed and currently marginalised groups, want from the state in these post-colonial contexts? To answer this question, we look at what political values citizens hold and how these are formed in the process of engaging with the state. We also consider how political values are formed through the experience of injustices in everyday life. If we are to understand what values are important to citizens, then we need to look beyond just moments constructed as 'crises' and turn our attention to the everyday. This volume draws together a number of chapters that consider the lived experiences of citizens as a starting point for understanding political values.

In conceptualising political values, we draw on ideas such as *nomos*, a normative universe where social and political behaviour is socially constructed, reflecting what society considers just (Cover 1982). We also draw on the concept of moral economy – where the idea of a 'legitimizing notion' frames what citizens feel is 'moral' or fair to expect from the state, reflecting what they value, both socially and politically.[3] These concepts help surface what political values individuals and groups in our case studies see as just, 'right' or 'moral'. We show how political values evolve with time and place, and in relation to different socially constructed identities including race, gender, class, caste, religion, age, ability, sexuality, etc. As a result of these complexities, this volume shows how political values are not fixed and may not be legible through the democratic system. To make political values more legible, several chapters explore narratives as a means of surfacing and articulating political values. These narratives can make visible what citizens expect from the state and what they consider just.

Second, we look at why and how citizens resist the state. We consider examples of visible and organised forms of protest as an important indicator of discontent with the state. We also consider less visible forms of resistance, reflecting complex histories and power relations, such as the persistence of racism, caste-based discrimination and gender inequalities. Finally, this volume asks what the nature of political values and their representation tell us about perceptions of what is just, and ultimately, about the evolution of post-colonial social contracts. Across the chapters, an analysis of how the social contract is being fractured is presented. In concluding this introduction we consider the implications of this fracturing for the prospects of social justice.

Studying the 'post-colonial'

We have chosen to focus on Brazil, India, South Africa and Zimbabwe. In drawing together these examples, we do not intend to present this as a representative sampling of post-colonial states. Rather, we intend for these cases to allow us to ground theorising about political values, narratives and resistance in specific experiences and examples. Each of these countries embody post-colonial contexts that had democratic political settlements, marking the formal end of colonial and authoritarian forms of governance. Through complex histories of subjugation and exploitation, colonialism is, however, embedded in social, economic and political systems at multiple levels (Mamdani 2018; Mbembe 2019; Nyamnjoh 2016). Colonialism, in 'both its forms and its substance, posited the issue of contingent human violence' (Mbembe 2001: 13).

This volume thus acknowledges the significance of decoloniality, which is understood as the process of naming, confronting and subverting persistent forms of colonialism (see Ashcroft et al. 2006). Decoloniality is relevant to this volume in several ways. Central to the formal shift in the political system away from authoritarian and colonial forms of governance was the idea that the state is no longer opposed to its citizens, and that new forms of engagement between citizens and political authority can exist. Under these new dispensations, 'subjects' became democratic 'citizens' (Mamdani 2018). However, in practice, the post-colonial state has not automatically equated to a decolonised state. As the chapters show, colonialism persists in many forms and at many levels, from how knowledge is constructed (Ngwenya and von Lieres, Chapter 2, this volume) to the ways that claims for justice can be delivered in courts (Shankar, Chapter 8, this volume). The chapters in this book look at both individual and systemic levels of analysis, in the context of the *long duree* of the state. They reflect on how individuals experience the ghosts of subjecthood while supposedly being democratic citizens. In practice, many are actually spectres of citizens. It is the institutional system embedded in, rather than cleanly breaking away from, centuries of colonial rule that perpetuates the 'spectre' of citizenship. While the states included here have not met the expectations of many of their citizens, the premise of new post-colonial democracies was to turn the marginalised into the empowered, where new voices would be heard and reflected in democratic governance.

While we focus on four post-colonial states, we also acknowledge that there is a common thread to the current mode of politics in the global South and North: a shift towards (or rise of) populist leaders, and an increased polarisation of political debates and social divides. In this political moment, public protest and mobilisation is also catalysed around particular versions of who should be included in the developmental project of the state (Povinelli 2011). This is mirrored by deepening economic inequalities globally and in many national contexts (Bourguignon et al. 2016), heightened by the global pandemic of COVID-19 (Fisher and Bubola 2020). Both the global North and South are experiencing frustrations with the limitations of the democratic system. Thus, while this volume includes cases drawn from post-colonial states, the issues of how democracy and justice intersect with decoloniality are relevant to wider challenges across many contexts.

The democratic and post-colonial state

Despite the persistence of distinctly colonial patterns of power within our case studies, there are nonetheless important changes in the shift to a democratic state. The rejection of colonialism manifests in democratic systems that support, in theory, norms and practices such as equality in voting, the rule of law, a Bill of Rights, competition and representation.

In each case, a new constitution, symbolising a new social contract, marked the shift from authoritarian governance to formal democracy. Together with each constitution came a new articulation of rights connected to demands for social justice. In South Africa, the Constitution and Bill of Rights were celebrated as a new form of justice in which all citizens and residents are equal and have the right to live a dignified life, as stipulated in the preamble to the South African Constitution (1996). This created a social contract based on a highly inclusive and broad set of human rights, that nonetheless operates over and within the existing inequalities of apartheid. Similarly, in Brazil, a series of progressive rights is encoded in the 1988 Constitution, including the right to dignity. The promises of a developmental state represented through these rights were the result of a long series of struggles by trade unions and other activists (see Schattan Coelho, this volume). Although there is an extensive bureaucracy within the state at different levels in Brazil that is committed to the promises of the 1988 Constitution, and an evolving array of social movements working with the state and contesting the process of how social rights are realised, this has not been enough to maintain the legitimacy of the Brazilian state in the face of entrenched inequalities and widespread violence (Braga and Purdy 2019).

In India, though progressive rights such as the right to education, food, health and shelter were in the non-justiciable Directive Principles section of India's Constitution, the Supreme Court of India made these justiciable through their judgements from the 1990s. The Court connected these rights to the fundamental right to life by using the words 'right to life with dignity'. In tandem, political parties piloted Bills through parliament to create laws that achieve this purpose. Despite this, the turn towards Hindu nationalism in India has quickly eroded the veneer of religious inclusion that cloaked the state in its current version (Chatterji et al. 2019).

While often breached by state security forces, the Zimbabwean Constitution of 2013 includes a Bill of Rights supporting the right to dignity, equality and non-discrimination among others (*Constitution* of *Zimbabwe* Amendment (No. 20) Act 2013). These rights, set out in law, reflect an explicit commitment by the state to deliver in terms of social justice. This state's approach to social justice and development has faced multiple economic and social challenges since independence, due in no small part to the autocratic leadership of the ruling ZANU-PF, particularly under Robert Mugabe's presidency. While the state did support a potentially transformative form of land expropriation and redistribution in the 2000s, this did not lead to improved social and economic livelihoods for the vast majority of largely poor Zimbabwean citizens (Chipika and Malaba 2016). Therefore, while there is an important democratic settlement in each country, reflected in progressive constitutions, the delivery of social justice is uneven.

The four countries in this volume have established patterns of democratic elections (with varying degrees of freedom and fairness) and distributions of power via political parties. In South Africa and Zimbabwe, long-standing liberation movements, the African National Congress (ANC) and ZANU (PF), respectively, won landslide victories and have maintained their electoral dominance since independence (Raftopoulos 2013). In India, the Indian National Congress party maintained a grip on electoral control at the national level from independence until 1977, and then again from 1980 to 1989, and then on and off for the next two decades. The party is now in disarray while the Bharatiya Janata Party (BJP), a Hindu nationalist party, has seized and maintained control since 2014 (Jayal 2019; Malji 2018). In Brazil, the picture is complex. The control of Brazil's political system has vacillated across the political spectrum after the 1988 Constitution, post-dictatorship (Schmitt 2000).

Participatory democracy and 'people's power'

Allied to the idea of a new social contract and a new democratic state that rejected the violence and subjectivity of colonialism was a move to deepen democracy by taking it closer to 'the people'. In all of our case studies (albeit in varying degrees), this is seen through the idea of participatory democracy. The rise of participatory democracy was in part a response to the limitations of post-colonial liberal democracy, and it was a key feature of the democratic procedural system in the early stages of democracy in Brazil, South Africa and India. Participatory democracy was mostly driven by political parties and social movements that wanted more engagement in everyday decision-making, as well as a deepened practice of co-constructed governance. For example, the origins of an influential participatory democratic institution, participatory budgeting, is tied to the rise of the Partido dos Trabalhadores (Worker's Party) in Porto Alegre, Brazil, and eventually spread internationally (Avritzer and Navarro 2003). Decentralisation of decision-making and participatory mechanisms that support citizen involvement in policy formulation between elections is also found in municipal wards across South Africa (Piper and Deacon 2009) and in parts of India, most notably Kerala (Heller 2013). In India, the Panchayati Raj Amendment Act in 1992 brought local self-government to the village level (Malik 2005). Participatory democracy in Zimbabwe has been contentious as its post-liberation period was characterised by a *de facto* one-party state which saw local and national governance structures replicating the ruling party structures. After a push for democratisation by civil society in the 1990s, Zimbabwe passed a constitution in 2013 that provides, in theory, for devolved governance and improved democratic participation (Mapuva 2019).

While forms of participatory governance exist across all four countries, there are fundamental differences in the scope and nature of participation and underlying understandings about the role of citizens. In Brazil, participatory democracy involves citizens setting the priorities for policies; while in South Africa, citizen participation often performs more of a monitoring and accountability role in relation to policies set by the state. In Zimbabwe, in instances where they do exist, citizen participation mechanisms are largely tokenistic. In India, efforts to increase the representation of the lowest castes and tribes and women in positions of political power (political decision-making) have had variable success across the country.

The nature and drivers underpinning the various transitions to democracy in the four cases in this volume have been the subject of extensive study by a range of authors, largely looking at democratic processes and historical norms and influences behind these transitions (Avritzer 2009; Friedman 2019; Sachikonye 2017). In each case, there was a clear shift, after colonisation, towards democratic citizenship through participatory democracy, which raises the question of what democratic citizens want and expect from the state.

Framing the debates: From political values to the social contract

In light of these political settlements, and the explicit move towards democracy in each country, what do democratic citizens want? To answer this question, we turn to a closer examination of the values and narratives reflecting what citizens perceive as just. We then look at how citizens express their expectations of the state through protest and other forms of resistance. We consider, based on detailed empirical investigation, how these ideas and practices reinforce or fracture the social contract.

Understanding political values and perceptions of justice: Narratives and resistance

Conceptualising and surfacing political values

Political values reflect what citizens see as important in terms of their expectations of the state, their own roles in democratic governance and what rights a new social contract should embody. Political values in turn can inform perceptions of justice. It is important to note that political values do not necessarily inherently align with issues of social justice. Political values can reproduce conservative or reactionary social orders as well as resist injustices. At the centre of this discussion is an intention to understand how values are formed and represented.

Several authors use moral economy to analyse political values and their relationship to perceptions of justice and social contracts. A moral economy framing assumes that acts of citizenship (Isin 2008) are not simply 'rebellions of the belly', but are reactions against the violation of a set of cultural conventions that exist in a particular moment in time. Moral economy shapes common expectations of what is socially just, as well as the principles that should constitute a political community (Thompson 1971; Scott 1977). Thompson and Scott both focus on the role of custom and community consensus when discussing the ways in which values manifest. For Thompson (1971), for example, food riots were triggered by grievances that operated within a popular consensus regarding what practices (baking, milling, etc.) were legitimate and illegitimate. These grievances and the allied consensus to protest were based on a 'consistent traditional' view of social norms. The riots reflected a 'vision of a better world' and the crowd's activities were shaped by a legitimizing notion: the 'men and women in the crowd were informed by the belief that they were defending traditional rights or customs; and, in general that they were supported by the wider consensus of the community' (Thompson 1971: 78). Here, political values are formed through legitimizing notions that are supported by a wider group.

The concept of moral economy can help surface the political values that underpin a shared understanding of what is perceived as 'just'. Values can, according to Siméant (2015: 171), also be formed and derived from living conditions that are 'marked by a preoccupation with subsistence, linked to the reciprocal expectations of elites and the people, which are at once pragmatic and normative, and which concerns the fair distribution of wealth and the responsibility of leaders'. For Siméant, the shared experience of living conditions highlights that values are not necessarily initiated in the domain of ideas. Instead, 'a considerable amount of what takes place in these matters arises from concretely lived experience and the practice of domination' (2015: 171).

If values are understood as socially constructed norms emerging from experiences, then it is also important to consider how these norms are woven into legal, political and economic systems. Robert Cover (1982) posits 'nomos' as a normative universe where 'right' and 'wrong', 'lawful' and 'unlawful' are created through narratives that locate values in practice and give them meaning. Legal rules and institutions thus interact with other cultural forces to produce legal meaning:

> A legal tradition is hence part and parcel of a complex normative world. The tradition includes not only a corpus juris, but also a language and a mythos – narratives in which the corpus juris is located by those whose wills act upon it. These myths establish the paradigms for behavior. They build relations between the normative and the material universe, between the constraints of reality and the demands of an ethic.
>
> (Cover 1982: 9)

The concept of nomos is useful for thinking about political values because it connects the normative to the concrete, for example, through law, as explored by Shankar (this volume). While nomos is articulated in relation to legal tradition, it can also be interpolated into arguments about the relationship between values and political settlements, through the social contract.

Knowing how and why political values are formed is key to understanding the nature and strength of the social contract in the post-colonial state. Yet making values legible in contexts of extreme marginalisation is not a straightforward proposition, as inequalities shape what is known and how knowledge is created (Tuhiwai Smith 2013). A number of chapters engage with the epistemological and hermeneutic challenges of how to recognise and describe political values in these contexts, especially through narratives.

Narratives and storytelling can play important roles in bringing the lived experiences of injustice into view. Stories demonstrate how people are simultaneously included and excluded in different moments, spaces and interactions – storytelling allows an articulation of these contradictory subjective and intersectional experiences (Wheeler et al. 2020). For example, the role of telling self-reflexive personal narratives can be significant from a psychological point of view. This type of storytelling allows participants to use the story as a form of organising events and affect in a self-narrative, rather than the articulation of a fixed linear narrative (Hartman 2019). This means that in remembering, telling and re-telling personal experiences and crafting these into stories, stories acquire a self-making quality (Habermas 2018). Telling stories in this way is, in part, a process of articulating one's own identity (Trees and Kellas 2009). This means that narratives can highlight the importance of embodied experiences of injustice, which are inscribed on bodies and in everyday life. These experiences are often missing from analyses of political systems.

Telling stories is a relational process, so it explicitly considers the relationship between the personal and the structural (Jackson 2013). Meaning is generated by how stories are heard and responded to by others (Arendt 2000), and these meanings can point to understandings of political values and justice. For example, research in South Africa shows the importance of exploring how individual experiences that emerge from stories can challenge structural inequalities such as racism (Sonn et al. 2013). A unique contribution of this volume is to foreground the role of narratives in understanding political values and ideas about justice. A number of the chapters speak of storytelling as a way of surfacing and altering what values are at play, how they gain support in political contestation and institutional responses and how they uncover the complex interplay of power in social and political life.

Reading political values in the post-colonial state

Drawing from these concepts, the chapters in this book test and expand thinking on political values from a broad range of perspectives. Wheeler (Chapter 1) questions whether values can be assumed to be known in contexts of exclusion. Using the example of violence and associated traumas, she shows how storytelling allows values to emerge and evolve in relation to personal experiences of injustice. This argument challenges the assumption that groups are able to make claims based on known and legible values. Instead, Wheeler argues that a process is required to allow values to be articulated and to gain collective support. She goes on to trace how shared values can become part of a collective narrative that implies recognition by the state and the position to make claims, using the example of how trauma can be used to draw the boundaries of the perceived social contract. Ngwenya and von Lieres argue, in Chapter 2, that Western ontologies and epistemologies are limiting our understanding of the representation of values formation, and that we must unlearn these in order to read 'resistant texts', which are essential for understanding the complexity of the social contract in the global South. The consequence of this incomplete acknowledgement of the nature of values is to fundamentally misread what is happening in protest and resistance.

For Cooper-Knock (Chapter 3), a value implicit in the ANC's developmental state at the start of democracy was that of 'patience'. Having faith in the promise of delivery was the political work assigned to citizens by the ANC during the political transition; a failure to practice the politics of patience was not just a political flaw, it was a moral failing. Alongside patience, the ANC government also embraced an ideological position that values citizens as both rights-bearing and as moral, responsible consumers. These (at times clashing) values formed the basis of the post-apartheid social contract.

Anciano (Chapter 4) demonstrates, through ethnographic work with fishers, that there are, at any one time, a range of 'legitimizing notions' informing the act of protest or the act of poaching. These include a perceived historical and cultural right to fish, a belief that past racial exclusion justifies current poaching and demand for fishing permits, the idea that fish is a social good and a desire to establish agency in a context where fishers feel helpless against the bureaucracy of the state. Multiple legitimizing notions may be supported by large groups, or one individual may identify with only a singular notion.

Price's contribution (Chapter 5) raises questions about what constitutes the public in post-colonial South Africa. She makes the case for online spaces as an important site for reading how values are made legible. She goes on to argue that the narratives that emerge in online spaces in relation to real-life protests in Cape Town are replicating and reinvigorating racist and discriminatory values. Price's contribution calls for a critical rethinking of the legitimacy of values within societies in light of histories of exclusion. Finally, Price analyses discourses of citizenship through online spaces to examine how the construction of values can be based on particular understandings of which narrative is legitimate and whose narratives count.

Dube and Schramm (Chapter 6) trace the nuance and complexity of how values around water as a common good are shaped by specific histories and experiences in Bulawayo, Zimbabwe. Their argument explores two distinct moments of contestation over water in the city, in which different understandings of water as a common good resulted in surfacing conflicting expectations about the respective roles of citizens and the authorities. This argument points to competing and divergent values, shaped by questions of who should be included in access to a common good, and on what basis.

Looking at how values are formed in relation to land by Adivasi[4] groups in India, Mohanty (Chapter 7) shows how this social good has material, cultural and epistemological connotations. Land symbolises family and community ties, encompasses a symbiosis between nature and wisdom and is a source of livelihood. These values justify the Adivasi's resistance to the encroachment on their land by private mining companies. Shankar's contribution (Chapter 8) starts with the violation of the rights of Adivasi in India by the state. She traces how a legal petition can be designed to appeal in particular ways to the cognitive process of dominant groups, including judges, so that a marginal group (the Adivasi in this case) can tell a powerful counter-story. Shankar shows how this counter-story confronts and reframes violence by police and other agents of the state with claims for justice. By engaging with the legal process, marginalised groups make sense of their trauma in political terms and inhabit a dynamic space of citizenship where the post-colonial democratic state provides the framework, but agency rests with the citizen.

For Schattan Coelho (Chapter 9), a historical analysis of health services in Brazil shows how authorities, health professionals and citizens built a shared understanding of what was fair in terms of the right to health. Through collectively developing a set of 'legitimizing notions', health became understood as a right for all citizens and a duty of the state, which should guarantee free access to quality services for health promotion and recovery. Despite contestations over the nature of health service provision, including how decisions are made and who is included and excluded from this service, Schattan Coelho argues that the universal value of health service as a right has become deeply embedded in the Brazilian context.

Finally, Waisbich (Chapter 10) shows how multiple political values are shaping conflicting expectations of foreign policy and development cooperation in post-colonial countries like Brazil, India and South Africa – countries that have recently emerged as 'rising powers in international development'. Using the case of the BRICS (Brazil, Russia, India, China and South Africa) and its New Development Bank (NDB), she shows, for example, an institutionalised transnational umbrella organisation of non-governmental organisations (NGOs) adopting a classic rights-based development approach and challenging the new bank to prioritise investments in social and environmentally sustainable infrastructure. While more radical groups of 'grassroots organisations' support the values of social and environmental justice, they also expand their critique of the new bank, demanding people-centred development, transparency and accountability as core values. There is no one homogenising view within civil society in one country, let alone across borders.

Emerging from the chapters are an important set of insights on how political values are formed and their significance in understanding what democratic citizens want. First, the chapters demonstrate that, in the context of post-colonial states, the values underpinning resistance are not necessarily fixed. Indeed, members of a conventionally understood 'community' may vehemently clash over values. Our cases show the need to question the idea that 'communities', such as those living in a shared geographic area, race groups or castes are uniform and represent one set of political values. These fixed categories do not hold up in the face of the complexity and heterogeneity of the experiences represented in this volume.

Second, the chapters show that, in the context of post-colonial states, political values are not always easily known. Several chapters in this volume explore different aspects of how narratives and storytelling can be used to articulate shared values as well as situate these values in relation to meanings of social justice. These chapters consider the meanings of narratives and why these are significant for bridging personal experience and collective actions; how they bring individual trauma into a collective and political experience, constructing new publics; and how they make sense of different forms of resistance. While values may not be fixed, and at times are not easily legible, once surfaced, they do point to important understandings of citizens' view of what is just.

What do citizens see as just?

Meanings of political values are intricately linked to perceptions of justice. Indeed, issues of justice arise repeatedly in relation to post-colonial states (Aning et al. 2018; Phillips 2011). Each chapter considers fundamental questions about what citizens want from the state and whether the state has delivered on these expectations. The chapters explore what can be considered just in relation to a perceived grievance with, or guarantee from, the state. Across the cases, there are some common features of these concerns around justice when conceived in the context of post-colonial states.

The redress (or lack thereof) of inequalities informs understandings of what is just. For example, Cooper-Knock details the pre-figurative (practices that reflect a future society) notions of justice embedded in a politics of patience, as citizens wait for the state to deliver on post-apartheid constitutional commitments of justice and development. Anciano and Dube and Schramm consider cases where there are claims for greater justice in how natural resources are accessed and controlled, in light of racial inequalities. Mohanty and Shankar explore the efforts of Adivasi groups to achieve justice in relation to the violation of their constitutional rights in India. Schattan Coelho addresses the just allocation of health resources, and Waisbich looks at the role of international collaboration on perceptions of justice.

A number of chapters highlight how rights and justice intersect in particular cases where lived experiences are often at odds with the rights set out in law. Anciano demonstrates how small-scale fishers may have the right to access fishing permits but in practice struggle to do so; Mohanty draws out the tensions between the right to land and the impact of the neoliberal market in denying the Adivasi this right. Shankar describes how the procedural rules of the legal system shape what kind of justice is possible and describes how, despite the procedural odds against them, marginalised groups can tell powerful counter-stories in their claim for justice.

Another feature of what is considered just, as seen through the case studies in this volume, pertains to the shift from subject to citizen and the prospect of ascribed identities not determining political futures. In the democratic state, a citizen who is Adivasi or of a low caste in India should have the same rights as non-Adivasi and upper-caste citizens. Similarly, in South Africa, rights are guaranteed regardless of race, religion, nationality, gender, age or other identity markers. This means that not only are the material consequences of rights important (such as the right to water, or to land), but also the universality of these rights is paramount. For example, Dube and Schramm trace how access to water is constructed as a common good in different ways in relation to colonial patterns of distribution and control with racial overtones. Again, the chapters show how the shift from subject to citizen plays out in reality through empirical detail, and the tensions that emerge.

There is a strong focus in this volume on the lived experiences of justice and injustice, and how this exposes contradictions between the promises of the post-colonial democratic state and reality. Despite the claim to universality of rights, when we look empirically at everyday experience, there are conflicting *nomos* around justice. For example, Wheeler and Price explore the dissonances between the promises of universal rights in South Africa and the daily injustices experienced by marginalised groups. In some cases, deep injustices are being perpetuated under the guise of the democratic state. For example, Shankar, Wheeler and Mohanty's contributions detail state violence towards marginalised groups. If, in the post-colonial democratic state, shared understandings of justice underpin the legitimacy of the social contract, then these dissonances around the experience of justice are significant. Together, they point to pressing questions about whether citizens have had their claims fulfilled.

Resistance

A deeper analysis of resistance is important for understanding whether citizens feel that their expectations for justice are being met. As we set out, there is substantial literature on participation and democracy in the global South, and a key contribution of this literature has been to broaden conceptions of democratic participation to include diverse forms of political engagement in participatory governance (see Cornwall and Coelho 2006). This volume engages with different registers of participation, including formal and informal forms of protest, resistance and collective action.

Literature on individual or civil society resistance often focuses on a performative notion of protest, in which protests are public and legible – thus, protests can be seen and their content known. Thompson (1971: 79), for example, in reference to moral economy, claimed that protest was in response to an outrage to moral assumptions as much as it was to actual deprivation – that was the usual cause of 'direct action' in his example of a food riot. Analyses of health care in Brazil by Schattan Coelho and land protests in India by Mohanty demonstrate how mobilisations and protest by a range of civil society actors has propelled demands for social justice, legitimised by the right to citizen-centred access to health care, land and livelihoods.

Visible and organised forms of protest are an important indicator of discontent, and of perceptions of injustice. Several chapters (Anciano, Mohanty, Price, Dube and Schramm, Shankar, Waisbich) describe visible social movements or community protests; however, they also show that the values underlying resistance and the ways they manifest are multiple and complex. For example, Waisbich explains how engagement in BRICS development planning initially focused on participation by civil society in invited spaces, with less critical input. This engagement soon included claimed 'people's spaces' focusing more on social justice concerns and including a stronger critique of BRICS policies. Their voices highlight a perceived broken promise of the post-colonial state. In their mode of engagement, one group prioritises (patient) participation, the other protest.

Current events in post-colonial states in this volume problematise the idea of a linear relationship between a protest and the demand for a social good or a right. While some of our cases highlight that citizens do protest for a social good or right, others show how protest has gone beyond this to include protesting to be seen or recognised (Dagnino 2005a). Protest can also be very ritualised and scripted (von Holdt et al. 2011). Additionally, protests can represent the rejection of social goods on the basis of how and to whom they are delivered by the state. Anciano shows how, on the one hand, small-scale fishers demand fishing permits and, on the other, will protest when they feel those permits are not given to 'legitimate' beneficiaries.

Therefore, the act of protesting and other forms of resistance are also important because they can constitute the boundaries of the groups involved. Through the act of protesting, certain groups can be included or excluded in making claims, and the labels applied to particular groups can be shifted. For example, Price explores the online/offline construction of publics through protest in relation to the historical categories of apartheid. She shows how an online space for a middle-class neighbourhood positions physical protests in a nearby and much poorer community as dangerous and illegitimate, reinforcing historical boundaries of whiteness and privilege. Shankar traces how an Adivasi group uses a counter-narrative to claim a position of visibility in the legal system, challenging the dominant narrative that positions them as criminals and threats to public order. Dube and Schramm explore the shifting political subjectivities of protesters in relation to national and local authorities in Bulawayo. They use the example of water as a common good to show how positionalities in protest actions are not fixed but emerge through a complex network of relations. Through the distinct tactics deployed, as well as the way in which issues of who has legitimacy in controlling access to water, Dube and Schramm argue for a more nuanced reading of how the groups involved shape protest and are in turn shaped by protest.

We argue that an understanding of resistance is incomplete if it focuses only on visible protest. It is necessary to also consider less visible forms of resistance. Beyond contestation, in the quieter moments of waiting for the state (Auyero 2012; Oldfield and Greyling 2015) and through forms of 'silent citizenship' (Ngwenya and von Lieres, Chapter 2, this volume), citizens can display a strong sense of what they feel is just, or a right. Ngwenya and von Lieres explore other interpretations of silence: when citizens are unable to act because of internalised oppression as well as when citizens are silenced by the state. They argue that the lack of protest cannot be assumed to indicate assent or inclusion.

In this volume, examples span visible and less visible forms of resistance to provide a more in-depth understanding of how citizens are responding to their perceptions of what is just and what they expect from the state. These different examples show that perceptions of values and justice are connected to resistance, but not in a linear way. Resistance may be an indication of citizens' discontent with the state, but it may also reflect other concerns such as the nature of claims or grievances, and the boundaries of the groups involved. A more subtle reading of resistance points towards reconsidering the relationship between citizens and the state. Contestation by citizens may not conform to expectations of liberal notions of participation in democracy and may be motivated by a complex mixture of material and moral imperatives, reflecting divergent understandings of the social contract.

The social contract and the post-colonial state

What do these examples of resistance, and the political values that inform resistance, tell us about the social contract between the post-colonial state and its citizens? This section will briefly outline the concept of a social contract, demonstrating how there can be multiple social contracts at different times and in different contexts, and how expectations aligned to these will shift over time. We then show that for the historically oppressed citizens in our cases, a social contract embodies the promise of a state to provide social justice. While aspects of these promises have been delivered, many shortcomings exist, leading to perceptions of a fracturing social contract.

Theorising on the social contract can be traced back to the ancient Greeks, the writings of Hobbes, Locke, Rousseau and, more recently, Rawls (Boucher and Kelly 1994). Arguably, whether in ancient Greece or ancient Africa, the governance of any society would require a stated or unstated set of norms that guided and structured interactions between the state (or rulers) and society. Understanding what these norms are in a post-colonial state can show what citizens expect from the state and what the state has promised to facilitate. Surfacing these norms is a difficult task, requiring an understanding of what political values citizens hold paramount. A contract between state and society reflects a normative world, based on narratives which in turn may be translated into codified paradigms, often embodied in a constitution (see Cover 1982). A social contract establishes expectations of behaviour by both state and citizen over a period of time. In the democratic system, an embedded norm, for example, is that citizens can change the government if they feel it is not delivering on the promises of a social contract.

Our cases show that the basis for political order, and its allied responsibilities and expectations, are created by varied and often competing social groups. Thus, there is never just one set of meanings attributed to the *nomos*, or one way in which a social contract is established or interpreted. It is important, therefore, not to assume a fixed and therefore always easily legible relationship between citizens and the state in the social contract. It is necessary to think beyond a relatively rigid notion of the 'equilibrium' of the social contract, which may imply that there are historic and/or fixed values. Instead, there are many examples of how multiple notions of the social contract co-exist, as well as how the understanding of the social contract moves with the particular moment in time (see Oldfield and Greyling 2015; Anciano and Piper 2019; Ndinga-Kanga et al. 2020). Challenges to a social contract can thus arise in part from different expectations of what is 'just' (Hossain and Kalita 2014).

An example of how the nature of a social contract can vary and how it can relate to different citizens at different times is reflected in Chatterjee's (2004) work on political and civil societies in India. He argues, using a Foucauldian perspective, that on the one hand, the social contract exists to govern populations that have to 'be both looked after and controlled by various governmental agencies' (Chatterjee 2004: 38). These members of 'political society' present themselves to the state to be looked after. This is different from the social contract envisaged at the moment of political settlement, after colonialism, where there was a legal coding of the new relation between the state and members of 'civil society': civil society would encompass the representation of rights-bearing citizens in a social contract with the state.

Just as our cases demonstrate the complexity of values embedded in a social contract, so too do they highlight that there is no singular version of 'the state', or one public authority in a social contract. Several chapters show that state or government officials are not the only authority in any given governance context. In post-colonial states, a complex range of public actors can wield authority – from state departments, to local brokers, to political parties. Ethnographic research on urban governance in the global South, for example, points to multiple and contending forms of governance, including bureaucratic (state-led), developmental (state-led), network (often civil society-led) and informal governance (Anciano and Piper 2019). Where there are multiple forms of governance, there are also numerous, intermeshed state and non-state public authorities. These authorities may be vested in state institutions, but as Lund (2006: 685) describes, they are often 'twilight institutions' mirroring the role the state theoretically plays. They mimic the state, but are at the same time able to enforce collectively binding decisions on members of society, who in most cases voluntarily acquiesce.

Anciano shows, for example, how leaders of fisher representative groups at different times mimic the authority of the state when they have to feed information back to their constituency, but at other times may willingly engage in and support protest against state officials. Waisbich examines different civil society actors trying to influence BRICS policy. One group, that Bond and Garcia (2015) call *BRICS-From-Above*, includes state officials, the private sector and government-affiliated think-tanks. This breaks the dichotomy of state and non-state actors. In a related point, Cooper-Knock discusses the multiple (and sometimes conflicting) logics that drive the discourse and actions of state representatives in everyday life (Joseph and Nugent 1994; Hansen and Stepputat 2001). As much as there are multiple public authorities, conflicting logics are often held simultaneously by the same state institution or official (Gupta and Sharma 2006).

While we have shown that political values and public authority may not be fixed, in a post-colonial context, a social contract implies some form of common political values. Our cases show that these are based on the key expectation that the state will redress historical and contemporary injustices. As we set out earlier, a new constitution encoded a new social contract and with each of these came a new articulation of rights connected to demands for social justice. Each country, to varying degrees, pledged to address racial and caste inequality, deal with social/historical discrimination, provide fairer access to livelihoods, deepen participatory governance and accountability, and defend dignity.

Fracturing social contracts?

If providing redress for social injustice is a key expectation that citizens have in relation to the social contract, it is clear from our case studies that this contract is fracturing. While the cases describe many advances made in support of justice, such as introducing participatory democracy and the provision of social welfare, for example, through extensive social grants in South Africa and Brazil, these have not been enough to address deeply embedded injustices. Several chapters address these points and show how the social contract is fracturing in response to persistent injustices.

Cooper-Knock shows how one form of a social contract was established in South Africa when the incoming ANC government created a moral economy of citizenship in 1994, which summoned citizens to adopt a politics of patience. This constitutional order provided citizens with strong socio-economic rights as well as political and civil protections; however, these rights were pre-figurative. These values may have been co-constructed in the anti-apartheid struggle, but in practice, they are promises that required citizens to be patient in their wait for freedom and dignity.

Anciano argues that the rise of protests and poaching in Cape Town demonstrates the manifestation of a fracturing social contract between the developmental promises of the 1996 South African Constitution and citizens who feel unable to sustain a livelihood. She shows the ways in which low-income fishers feel betrayed by the post-transition allocation of fishing permits and how this reflects a failure of the state to uphold its developmental promises, echoing Cooper-Knock's view of a pre-figurative social contract.

Dube and Schramm, in exploring the interplay between contestation over water as a common good and the repertoires of protest in Bulawayo, demonstrate how colonial histories and the specific politics of the city shape the way that protests unfold – revealing shifting assumptions about the social contract. In their chapter and in the case of the water protests, neither the legitimacy of the national state nor that of the local government was a given. The legitimacy of the social contract rested on the respective authority's capability to provide adequate services. The protesters also framed their claims primarily in relation to the city and not the nation, appealing to very specific, even personal, relations and responsibilities. This example shows how the frame of the social contract shifts in specific examples.

Mohanty argues that democracy and development are the twin pillars on which the social contract between the state and citizens is founded in post-independence India. This is encapsulated by a political regime founded on the democratic liberal principles of rights and a welfare system that promises to support the socially and economically marginalised. In this vein, the Adivasi understanding of their right to land is based on a social contract that is mutually constructed and has evolved over decades following independence from colonial rule. This contract was thus breached by the state when it diverted Adivasi land to mining corporations. The state has failed to uphold its contract because, she argues, it has retreated; it has shifted from the position of a patron and an agency of welfare to a position of facilitator of neoliberal capital. While a normative grounding that sets a transformative agenda for the state has not yet been completely abandoned, it has significantly lost its power to honour the social contract it has with the poor.

Schattan Coelho argues that the 1988 Brazilian Constitution formalized a new social contract between citizens and the state, in which the legitimacy of the state was linked to its capacity to assure the development of citizens' capabilities – particularly in relation to the right to health services. New political leadership in the form of a right-wing, conservative President may undermine this contract, leading, Schattan Coelho predicts, to growing social protest if progressive forms of health provision are eroded.

Our case studies raise a series of important issues. First is the extent to which these social contracts, in their moments of conception and future iterations, reflect an elite hegemonic project or a co-constructed developmental agenda. Second, in the context of post-colonial states, the temporality of this social contract is paramount. Cooper-Knock highlights the importance of this when she shows that the social contract embraces the pre-figurative idea of social justice, rather than a historically established and embedded social contract. Political struggles that want to preserve what has already existed are fundamentally different from those that hope to realise what has not yet materialised. In cases where both are perceived to be fractured, however, the social contract is damaged. Third, to whom do citizens turn to uphold the social contract, in contexts where there are multiple public authorities? Anciano shows, for example, that understanding the fractured post-apartheid social contract requires an analysis of public authority. Where there are multiple public authorities (including the state, community representatives and political parties), all of them hold authority to maintain the social contract in order to deliver a form of social justice. Where there are multiple governance actors, where do citizens turn in their claims for social justice?

Conclusion

This volume offers an interdisciplinary and multi-scalar approach to the analysis of pressing political, social and economic issues in four post-colonial states. What has emerged is a complex picture that at times challenges and extends some of the central framings within the existing literature, particularly in the deepening democracy debates. The strong focus in this volume on the lived experiences of justice and injustice draws attention to what citizens want in the post-colonial state, and the political values that emerge from their experiences.

This collection makes the case for a deeper and wider understanding of political values to understand contemporary challenges for democracy in post-colonial contexts. This is because democratic political systems do not always represent sufficiently the breath of political values in diverse contexts. Chapters such as those by Wheeler, Shankar, Ngwenya and von Lieres point to the importance of surfacing values through everyday experiences and anti-colonial ways of knowing. Participatory democracy, while an improvement on narrower forms of democracy in terms of inclusion, still does not allow for a full reflection and representation of political values that may be historically silenced (Lemanski 2017). Protest and other forms of resistance, while important, also do not necessarily reflect the breadth of relevant political values. This is partly because visible protest is not the only form of resistance that we should consider (a number of chapters make the case for looking at more diverse forms of resistance, including forms of silence). The systemic exclusions embedded into the democratic system, from participation to protest, prevent some groups that have been historically marginalised from being represented. We have shown how a post-colonial democratic social contract does not automatically translate into the inclusion of perspectives that have been historically silenced. Without a fuller representation of these values, across the whole spectrum of historically and currently marginalised citizens, citizens will continue to feel that their claims for justice are not being heard and met, which puts pressure on the social contract. If enough pressure is created on the social contract, there are questions as to how long it can be sustained.

When the social contract breaks down, there are serious consequences: it can lead to the government being seen as illegitimate, resulting in low-level conflict and violence, and in some cases, even civil war. However, there is not (yet) a complete breakdown of the social contract in the countries in this volume. Instead, we have shown that there can be multiple versions of the social contract in operation. This suggests that the loss of the legitimacy of the social contract is not a binary: there can be varying degrees of legitimacy to the social contract and different interpretations of it. Ultimately, questions about the social contract come down to how the post-colonial state is able to meet citizens' expectations for justice.

While the social contract is not broken, the disconnect between political values and perceptions of what is just is leading to its fracturing. Without a more nuanced understanding of the challenges to the social contract, it is not clear how these fractures will be addressed. The chapters in this volume suggest some possible ways forward. First, new narratives can be created that bridge gaps in political values and perceptions of justice, as in Shankar's chapter. Crucially, it is not the political process that has made this bridge: it is the reframing of dominant narratives through a counter-story that leads to its inclusion in the justice and political system. This dynamic is part of the prominence of the Black Lives Matter movement. New narratives and counter-narratives hold the potential to provide systematic social justice. Second, some cases show that when political values are not being heard, there are civil society organisations that will amplify these voices and fight for social justice for the historically marginalised. These organisations can put pressure on the other parts of civil society, the government, ruling parties and the state to amend the social contract – preventing conflict and the fracturing of the state.

As we conclude this volume, many countries are experiencing profound challenges to democracy. In the United States, the social contract is under extreme pressure in new ways, leading to constitutional crises. In the countries considered in this volume, the global pandemic of COVID-19 is shining light on the existing structural inequalities and systemic weaknesses in governance. In both older and newer democracies, the questions of what citizens want and how states can deliver on their promises are more salient than ever. Therefore, the questions raised in this book are ones that scholars can return to over time. We need to continuously engage with questions of how citizens understand social justice to meet the challenges of post-colonial democracy.

Approach, methodology and structure

Approach

Across the chapters in this volume, there are multiple levels of analysis, providing insights into the international, national and local levels as well as individual perspectives on the relationship between citizens and the state. The chapters connect the micro-level experiences of living with injustice and immediate livelihood challenges, while also providing an avenue for analysing structural dynamics. While authors in the book investigate resistance and political values through linking varying levels of analysis and phenomena, the chapters are, at the same time, grounded in extensive empirical research. These chapters have been written using different registers of analysis of political systems and structures, with some including rich empirical detail from ethnographic and participatory research.

Several chapters focus on national settings. Cooper-Knock unpacks citizenship and resistance in South Africa, Schattan Coelho looks at Brazil's national health system and Mohanty turns to the sub-national level, interrogating land protests in relation to the Adivasi (indigenous people). Anciano and Dube and Schramm look at moral economies of protest in urban settings through specific groups and communities in Cape Town and Bulawayo respectively. Ngwenya and von Lieres address the level of the individual, including the psyche, and its interface with the politics of protest at the group level. Shankar looks at how individual and group narratives play out in the context of a petition before the higher judiciary in India. Price considers the narratives of groups through online spaces. Wheeler too examines personal experiences of injustice and violence through storytelling, and how values emerge through articulating personal experiences that lead to collective political positions.

This book is also deliberately multidisciplinary. It includes analyses from the fields of political studies, anthropology, development studies, human geography, urban planning, legal studies, history, sociology and psychology. Insights from different disciplines allow for a more nuanced and holistic understanding of the central questions. To unpack the layers of political values in different forms of resistance, and to understand conceptions of social justice, we encouraged authors to use a range of theoretical constructs. The concepts that emerged as useful were moral economy, citizenship, social justice, *nomos* and common good. These are all concepts that can bridge micro and macro levels of analysis. All of them also encourage a focus on the intersection between political values and practices of resistance. For example, moral economy speaks to the key themes of values formation, what is perceived as legitimate, which actors hold public authority and what this tells us about social contracts.

Methodology

Methodologically, the book draws on diverse disciplinary approaches: from ethnography to storytelling to state-level political analysis. It includes a range of methods informed by the specific questions and contexts involved in each chapter. All chapters have a strong empirical grounding explored through qualitative, quantitative and participatory methodologies depending on each case.

As a volume, it has been developed collaboratively. This book and our approach emerges out of the Collaboration for Research on Democracy (CORD), which brings together scholars largely from the global South, working through grounded and interdisciplinary research. The volume both builds on collaborative work done by CORD, including scholars who have been working together over many years, as well as early career researchers based in the global South. Many of the key ideas and chapters contained in this volume were developed collaboratively at CORD workshops held in Bellagio, Italy, in March 2017; and in Cape Town, South Africa, in July 2018.[5] These workshops provided the basis for the intellectual project that underpins this volume, in part through bringing together scholars, activists and practitioners concerned with social justice. The book focuses on challenging the dominance of the global North in the framing of the questions it deals with. The editors employed mentoring and supportive editing to develop each chapter. The introduction was written inductively based on what emerged from the empirical chapters and with significant discussion and input from all authors. This book offers fresh contributions to thinking on political values formation, resistance and the social contract. It brings together grounded scholarship from different disciplinary perspectives to provide new and critical empirical insights and draw out theoretical implications. It also brings together younger scholars and more experienced academics, largely based in the global South, drawing on extensive and in-depth field work and engagement in the global South. In a field largely dominated by scholars and work developed in the global North, these perspectives are important as a counterpoint in relation to wider academic debates.

Structure of the book

The book chapters take us on a journey, from individual narratives of injustice to international analysis of political values underpinning resistance to foreign policy. We move from South Africa to Zimbabwe, to India and finally Brazil. We start with Wheeler, in South Africa, who traces the formation of political values through personal and collective storytelling, using the example of trauma to examine the implications for understandings of justice. Ngwenya and von Lieres focus on different readings of silences and the forms of knowledge beyond those generally recognised within Western ontologies and epistemologies. They argue for the need to unlearn Western-centric approaches to access stories and interpret forms of silence, which are often missed or distorted within dominant political systems. Cooper-Knock looks at how moral economy as a concept is linked to the idea of citizenship in the context of post-apartheid South Africa. Anciano reflects on the social contract in contemporary South Africa, looking at political values underpinning protest and poaching in a suburb of Cape Town, and what they tell us about the relationships between citizens and public authorities. Price situates her chapter in relation to protests in contemporary South Africa and explores how the creation of online narratives about citizenship and belonging are used to promulgate exclusionary values. Moving to Zimbabwe, Dube and Schramm trace the evolution of two distinct periods of contestation over water in Bulawayo, and how water is constructed differently as a common good. Mohanty takes us to India, where she looks at how the state has breached the social contract with the Adivasi by supporting large-scale mining on Adivasi land. Staying in India, Shankar addresses the complex problem of how the validity of counter-stories of the marginalised can be established, how these counter-stories can use cognitive processes to create bridges or overlaps between different nomos, and how engaging with the narrative framework of the law assists the marginalised to make sense of their trauma in political terms and demand justice within the framework of constitutional rights. Schattan Coelho examines political values and demands for inclusion and social justice in the public health system in Brazil. She looks at the historical evolution of the health system and what this tells us about the social contract in Brazil. Waisbich looks at the formation and contestation of political values through the lens of the BRICS NDB. These thought-provoking chapters bring to the fore how political values emerge, through narratives and resistances, pointing towards claims for social justice in post-colonial states.

Notes

1 We have consciously used the term 'post-colonial state' in our title rather than the term 'global South' as we feel that the latter is problematic, and can be understood to reinforce certain elements of colonialism. There are ways in which the category of 'global South' enables the perpetuation of colonial tropes in scholarship, aid, and politics. While we are cognizant of the problematic nature of the 'global South' as a research category, we are nonetheless a group of researchers working largely in post-colonial countries in the 'global South'. Thus, we recognise the importance of giving greater attention to the realities in places beyond the dominant references in Western academia. Where the term 'global South' is used in this book, it is used with this reflection in mind.

2 Across the volume, we understand citizenship to be inclusive and not limited to those with legal status (Kabeer 2005). Citizenship has a normative component that recognises the right to have rights (see Dagnino 2005b) and can be relevant across different scales and territories (Yiftachel 2015; Yuval-Davis 2007).

3 For a more detailed discussion of nomos and moral economy, see the section on conceptualising and surfacing political values in this chapter as well as Cooper-Knock, Anciano and Shankar (Chapters 3, 4 and 8, this volume).

4 The literal meaning of Adivasi is 'forest dweller'. Adivasi are a Scheduled Tribe in India's legal language. See Mohanty and Shankar (Chapters 7 and 8, this volume).

5 Funding was generously provided by the National Institute for Humanities and Social Sciences in South Africa, and the Rockefeller Foundation in Italy supported these workshops and the development of this volume.

References

Anciano, F., and L. Piper. 2019. *Democracy Disconnected: Participation and Governance in a City of the South*. London: Routledge.

Aning, K., M.A. Brown, V. Boege and C.T. Hunt. 2018. *Exploring Peace Formation: Security and Justice in Post-Colonial States*. London: Routledge.

Arendt, H. 2000. *Stateless Persons: The Portable Hannah Arendt*. New York: Penguin.

Ashcroft, B., G. Griffiths and H. Tiffin (eds.). 2006. *The Post-Colonial Studies Reader*. Oxford: Taylor & Francis.

Auyero, J. 2012. *Patients of the State: The Politics of Waiting in Argentina*. Durham, NC and London: Duke University Press.

Avritzer, L. 2009. *Democracy and the Public Space in Latin America*. Princeton, NJ: Princeton University Press.

Avritzer, L., and Z. Navarro. 2003. *A inovação democrática no Brasil: O orçamento participativo*. São Paulo: Cortez Editora.

Bond, P., and A. Garcia. 2015. *BRICS: An Anti-Capitalist Critique*. London: Pluto Press.

Boucher, D., and P. Kelly. 1994. *The Social Contract from Hobbes to Rawls*. London: Routledge.

Bourguignon, F., et al. 2016. 'Inequalities: Many Intersecting Dimensions', in ISSC, IDS and UNESCO (eds.), *World Social Science Report 2016: Challenging Inequalities – Pathways to a Just World*. Paris: UNESCO Publishing, pp. 41–82.

Braga, R., and S. Purdy. 2019. 'A Precarious Hegemony: Neo-Liberalism, Social Struggles, and the End of Lulismo in Brazil', *Globalizations* 16(2): 201–215.

Chatterjee, P. 2004. *The Politics of the Governed: Reflections on Popular Politics in Most of the World*. New York: Columbia University Press.

Chatterji, A.P., T.B. Hansen and C. Jaffrelot. 2019. *Majoritarian State: How Hindu Nationalism Is Changing India*. New York: Oxford University Press.

Chipika, J.T., and J.A. Malaba. 2016. 'Towards a Transformative Democratic Developmental State in Zimbabwe', in G. Kanyenze and H. Jauch (eds.), *Towards Democratic Development States in Southern Africa*. Zimbabwe: Weaver Press.

Constitution *of* Zimbabwe Amendment (No. 20) Act. 2013. *Zimbabwe*, 22 May.

Cornwall, A., and V. Coelho. 2006. *Spaces for Change? The Politics of Citizen Participation in New Democratic Arenas*. London: Zed Books.

Cover, R. 1982. 'Nomos and Narrative', *Harvard Law Review* 97(4): 4–64.

Dagnino, E. 2005a. 'Meanings of Citizenship in Latin America', *IDS Working Paper*. Brighton: Institute of Development Studies.

Dagnino, E. 2005b. '"We All Have Rights, But …": Contesting Concepts of Citizenship in Brazil', in N. Kabeer (ed.), *Inclusive Citizenship: Meanings and Expressions*. London: Zed Books, pp. 149–156.

Fisher, M., and E. Bubola. 2020. 'As Coronavirus Deepens Inequality, Inequality Worsens Its Spread'. *New York Times*. Accessed July 23 2020. https://www.nytimes.com/2020/03/15/world/europe/coronavirus-inequality.html.

Friedman, S. 2019. *Power in Action: Democracy, Citizenship and Social Justice*. Johannesburg: Wits University Press.

Gaventa, J., and R. McGee. 2010. 'Introduction: Making Change Happen – Citizen Action and National Policy Reform', in J. Gaventa and R. McGee (eds.), *Citizen Action and National Policy Reform: Making Change Happen*. London: Zed Books, pp. 1–43.

Gupta, A., and A. Sharma. 2006. 'Introduction: Rethinking Theories of the State in an Age of Globalisation', in A. Gupta and A. Sharma (eds.), *Anthropology of the State: A Reader*. Oxford: Blackwell Publishers, pp. 1–42.

Habermas, T. 2018. *Emotion and Narrative: Perspectives in Autobiographical Storytelling*. Cambridge: Cambridge University Press.

Hansen, T., and F. Stepputat. 2001. 'Introduction: States of Imagination', in T. Hansen and F. Stepputat (eds.), *States of Imagination*. New York: Dukes Press, pp. 1–40.

Hartman, S. 2019. *Wayward Lives, Beautiful Experiments: Intimate Histories of Riotous Black Girls, Troublesome Women, and Queer Radicals*. New York: W.W. Norton & Company.

Heller, P. 2013. 'Participation and Democratic Transformation: Building Effective Citizenship in Brazil, India and South Africa', in K. Stokke and O. Törnquist (eds.), *Democratization in the Global South: The Importance of Transformative Politics*. London: Palgrave Macmillan, pp. 42–74.

Hossain, N., and D. Kalita. 2014. 'Moral Economy in a Global Era: The Politics of Provisions during Contemporary Food Price Spikes', *Journal of Peasant Studies* 41(5): 815–831.

Isin, E. 2008. 'Theorizing Acts of Citizenship', in E. Isin and G.M. Nielsen (eds.), *Acts of Citizenship*. London: Palgrave Macmillan, pp. 15–43.

Jackson, M. 2013. *The Politics of Storytelling: Variations on a Theme by Hannah Arendt*. Copenhagen: Museum Tusculanum Press.

Jayal, N.G. 2019. 'Reconfiguring Citizenship in Contemporary India', *South Asia: Journal of South Asian Studies* 42(1): 33–50.

Joseph, G., and D. Nugent. 1994. *Everyday Forms of State Formation: Revolution and the Negotiation of Rule in Modern Mexico*. Durham, NC: Duke University Press.

Kabeer, N. 2005. *Inclusive Citizenship: Meanings and Expressions*. London: Zed Books.

Lemanski, C. 2017. 'Unequal Citizenship in Unequal Cities: Participatory Urban Governance in Contemporary South Africa', *International Development Planning Review* 39(1): 15–35.

Lund, C. 2006. 'Twilight Institutions: Public Authority and Local Politics in Africa', *Development and Change* 37(4): 685–705.

Malik, A.S. 2005. 'Local Self-Government at Village Level—An Assessment', *The Indian Journal of Political Science* 66(4): 773–792.

Malji, A. 2018. 'The Rise of Hindu Nationalism and Its Regional and Global Ramifications', *Education About ASIA* 23(1): 39–43.

Mamdani, M. 2018. *Citizen and Subject: Contemporary Africa and the Legacy of Late Colonialism*. Princeton, NJ: Princeton University Press.

Mapuva, J. 2019. 'Increasing Participatory Space in Zimbabwean Local Governance Democracy', *African Journal of History and Culture* 11(7): 65–73.

Mbembe, A. 2001. *On the Postcolony*. Berkeley, CA: University of California Press.

Mbembe, A. 2019. *Necropolitics*. Durham, NC: Duke University Press.

Ndinga-Kanga, M., H. van der Merwe and D. Hartford. 2020. 'Forging a Resilient Social Contract in South Africa: States and Societies Sustaining Peace in the Post-Apartheid Era', *Journal of Intervention and Statebuilding* 14(1): 22–41.

Nyamnjoh, F.B. 2016. *#RhodesMustFall: Nibbling at Resilient Colonialism in South Africa*. Cameroon: Langaa Research and Publishing.

Oldfield, S., and S. Greyling. 2015. 'Waiting for the State: A Politics of Housing in South Africa', *Environment and Planning A* 47(5): 1100–1112.

Phillips, R. 2011. 'Postcolonial Scholarship in Social Justice Research', in L. Markauskaite, P. Freebody and J. Irwin (eds.), *Methodological Choice and Design*. Dordrecht: Springer, pp. 157–166.

Piper, L., and Deacon, R. 2009. 'Too Dependent to Participate: Ward Committees and Local Democratisation in South Africa', *Local Government Studies* 35(4): 415–433.

Povinelli, E. 2011. *Economies of Abandonment: Social Belonging and Endurance in Late Liberalism*. Durham, NC: Duke University Press.

Raftopoulos, B. 2013. 'The 2013 Elections in Zimbabwe: The End of an Era', *Journal of Southern African Studies* 39(4): 971–988.

Republic of South Africa. 1996. *Constitution of South Africa 1996*. Pretoria: Republic of South Africa.

Sachikonye, L. 2017. 'The Protracted Democratic Transition in Zimbabwe', *Taiwan Journal of Democracy* 13(1): 117–136.

Schmitt, R. 2000. *Partidos Políticos no Brasil, 1945–2000*. São Paulo: Zahar.

Scott, J.C. 1977. *The Moral Economy of the Peasant: Rebellion and Subsistence in Southeast Asia*. London: Yale University Press.

Siméant, J. 2015. 'Three Bodies of Moral Economy: The Diffusion of a Concept', *Journal of Global Ethics* 11(2): 163–175.

Sonn, C.C., G. Stevens and N. Duncan. 2013. 'Decolonisation, Critical Methodologies and Why Stories Matter', in G. Stevens, N. Duncan and D. Hook (eds.), *Race, Memory and the Apartheid Archive*. London: Palgrave Macmillan, pp. 295–314.

Thompson, E.P. 1971. 'The Moral Economy of the English Crowd in the Eighteenth Century', *Past and Present* 50: 76–136.

Thompson, L., and C. Tapscott. 2010. *Citizenship and Social Movements: Perspectives from the Global South*. London: Zed Books.

Trees, A., and J.K. Kellas. 2009. 'Telling Tales: Enacting Family Relationships in Joint Storytelling about Difficult Family Experiences', *Western Journal of Communication* 73(1): 91–111.

Tuhiwai Smith, L.T. 2013. *Decolonizing Methodologies: Research and Indigenous Peoples*. London: Zed Books.

Von Holdt, K., et al. 2011. *The Smoke that Calls: Insurgent Citizenship, Collective Violence and the Struggle for a Place in the New South Africa*. Johannesburg: Centre for the Study of Violence and Reconciliation (CSVR).

Wheeler, J., J. Shaw and J. Howard. 2020. 'Politics and Practices of Inclusion: Intersectional Participatory Action Research', *Community Development Journal* 55(1): 45–63.

Yiftachel, O. 2015. 'Epilogue – From "Gray Space" to Equal "Metrozenship"? Reflections on Urban Citizenship', *International Journal of Urban and Regional Research* 39(4): 726–737.

Yuval-Davis, N. 2007. 'Intersectionality, Citizenship and Contemporary Politics of Belonging', *Critical Review of International Social and Political Philosophy* 10(5): 561–574.

1 Surfacing political values

Narratives of justice in Cape Town, South Africa

Joanna Wheeler

Introduction

Delft is a settlement that emerged after apartheid ended in South Africa in 1994, and was one of the first townships in Cape Town to be explicitly 'mixed-race', in line with the historic racial categories of apartheid.[1] Positioned on the sandy, wind-blown Cape Flats, thirty kilometres from the city centre, Delft reflects the complexity of transformation and social justice in South Africa: daily life in Delft is characterised by multiple forms of violence, an acute lack of jobs (especially for young people), the twinned rise of substance abuse and gangs, endemic corruption in the police and other state authorities, and local political leaders exploiting racialised histories for populist power (see Piper and Wheeler 2016 for similar examples from other settlements). Delft reflects not only the current challenges of social transformation in South Africa, but the global trends of urban hinterlands (Wacquant 2008) where the configurations of political authority and inclusion do not conform to Western conventions of democracy (Anciano and Piper 2019). Daily life reflects the intersections of multiple inequalities, as well as historic and contemporary trauma (Burns et al. 2013; Kabeer 2016). Within this context, understandings of social justice and injustice cannot be taken as given or known.

This chapter draws on participatory research conducted between 2015 and 2017 with a group of residents from Delft, Cape Town. This group included young people and older community activists, speaking a mixture of Afrikaans, isiXhosa and isiZulu.[2] The group members joined the research process because they were all concerned with issues of safety and security in Delft. Some were involved in neighbourhood watches and the Community Policing Forum,[3] and all were affected by the multiple forms of violence and lack of accountability within Delft. After an initial introduction via Sustainable Livelihoods Foundation, a local NGO that had worked in Delft on several different projects, I held an all-day workshop at a civic hall to explore the possibility of taking the research process forward. Around thirty people attended this workshop. We watched some personal stories told through other research projects and had an extended discussion about the storytelling approach and what it entails. From this workshop, twelve people joined the research process and chose the name 'Delft Safety Group' for themselves during the course of the storytelling process.[4]

The research process involved personal storytelling about everyday experiences of insecurity and safety in Delft, leading to a process of collective analysis of their own stories. The research group then used this collective analysis to identify themes for filmmaking, moving the narrative outwards to speak to the wider community and political system. Finally, the research involved a series of public events, where participants engaged different representatives of the state in dialogue about their demands and obligations, attempting to assert claims for justice based on their understanding of the social contract.

Within this broader research process, this chapter focuses on personal and collective storytelling and how the storytelling process surfaced and articulated shared values within a group. In addition to exploring how shared values are formed and become known through storytelling, it also examines how these values shape the expectations and positioning between citizens and the state through particular understandings of subjectivity and justice. It uses the example of trauma to show how values emerge in relation to making claims for justice.

Political values and justice

We argue in the Introduction to this volume that political values 'reflect what citizens see as important in their expectations of the state, their own roles in democratic governance and what rights a new social contract should embody' (Anciano and Wheeler, this volume). Yet the centrality of shared political values to claims for justice in the post-colonial context does not suggest that there are definitive or unified versions of political values. As Shankar argues in this volume with relation to the counter-narratives of marginalized groups in India, competing sets of values are often at play in the post-colonial context. The visibility of these different values is a reflection of persistent colonial power structures that subordinate certain types of experiences and particular groups (for example, Sonn et al. 2013). For every example of enfranchisement and substantive citizenship in contemporary South Africa, there is a counter example of how people have been either formally or informally excluded by the boundaries of citizenship or by the demarcation of rights (for other examples see in this volume Anciano, Chapter 4; Price, Chapter 5; Dube and Schramm, Chapter 6; and Mohanty, Chapter 7). Careful attention needs to be paid to the appropriate epistemologies to surface different forms of knowledge (Nyamnjoh 2012) and to recognize resistant texts (Ngwenya and von Lieres, Chapter 2, this volume). This chapter explores how personal and collective storytelling can offer an epistemological and heuristic lens for reading the political values that underpin normative assumptions about justice. The focus of analysis in this chapter is on how these values emerge through the *small* stories of the everyday (Bamberg and Georgakopoulou 2008). In this sense, storytelling is positioned as a form of intersectional inquiry (Wheeler et al. 2018). This approach recognises the multiple and intersecting ways that oppression is experienced across socially constructed identities such as race, class, gender, sexuality, ethnicity, ability, religion, etc. (Collins and Bilge 2016). As human beings, people are simultaneously included and excluded in different moments, spaces and interactions; and storytelling allows an articulation of these contradictory experiences more fully by bringing together affect and events into a coherent narrative (Wheeler 2018). This chapter will trace how political values can be articulated through this process.

A second aspect of storytelling and political values explored in this chapter is how collective expectations of how society should function, based on relations between the state and citizens, and what claims for justice can be expected within the social contract (Boucher and Kelly 1994). The social contract implies a shared narrative that shapes an imagined community in the post-colonial context (Anderson 2006). For example, discourses of citizenship in South Africa emerge from the hopes attached to the end of apartheid and the transition to democracy (see Cooper-Knock, Chapter 3, this volume). Yet it is not clear how this shared narrative originates, replicates and evolves, especially in contexts of entrenched inequality and social division. While further literature focuses on describing why protest and resistance as forms of political engagement happen, the focus of this chapter is on exposing the relationship between the body and the lived experiences that inform particular values, assumptions and collective identities for claim-making, and ultimately the nature of those claims.

The chapter goes on to explore how values surfaced through storytelling about personal experiences can lead to collective narratives in which claims to justice are made. These claims for justice are grounded in the values explored through personal stories. It then moves on to consider what these values suggest about claims for justice, using trauma as an example of how boundaries of justice are constructed. Therefore, this chapter offers a critique of the assumption that values are easily legible and known, based on a careful reading of stories and storytelling in the context of the case of the group from Delft. It argues that stories, personal and collective, offer a way of seeing and shifting political values in relation to everyday experiences that may not be recognised or articulated in the public domain.

Stories and the politics of storytelling

Stories are told in multiple ways and forms, in varying levels of depth and complexity. In *Aspects of the Novel*, E.M. Forster (1927 [1955]) argues that the story is a chronological sequence of events and the plot is the causal and logical structure which connects these events. This is a rational and distinctly Western view of story. Other definitions of story, more or less bounded by chronology, conceptions of action, plot and characters expand the frame further (for example, Emberley 2014; Zepeda 2014). Michael Jackson, drawing on Arendt, argues that storytelling can be defined as the means 'whereby a person, from early infancy, organizes and manages the life experiences that threaten to engulf, undermine, or nullify him or her' (Jackson 2013: 36). Stories can thus be distinguished from descriptions (which lack a plot) or even anecdotes (which may lack a story), which are common in qualitative research. The focus in this chapter is on the autopoesis of storytelling, the self-making quality of telling stories about one's own experiences (Wheeler and Bivens 2021). This analysis takes story as a form of organising events and affect in a self-narrative (Hartman 2019). An essential aspect of this focus, in terms of research, is that storytellers choose the event(s) and articulate the particular sequence based on their understanding of the plot (why things happen in a particular order). The next section explores the principles underpinning the methodological approach to storytelling and why this is relevant for understanding politics and justice.

Storytelling as politics: A methodology

The research in Delft employed a specific approach to storytelling developed by Wheeler and collaborators over fifteen years. A more in-depth exploration of this approach can be found in Wheeler, Shahrokh and Derakhshani (2018). It uses storytelling as a form of inquiry, through the exploration of the self and daily experiences situated within wider life histories and social context, using multi-layered and creative techniques. This leads to personal stories that are embodied and multi-modal (Reavey 2011). It includes the body as part of the semiotic repertoire, in what is termed 'translanguaging in multimodal communication' (García and Li 2014: 1), or finding a way to communicate the unsayable. This aspect of storytelling is important for understanding perceptions of values and justice that are related to experiences that are not easily expressed.

A group process enables participants to surface and articulate stories of personal and/or collective experience. It uses elements of story structure and storytelling that enable freedom in self-expression. It relies on iteration, between versions of stories and modes of expressions, and this iteration allows people to process significant moments in their lives, in turn assigning sequences and meaning to experiences. This iterative process of reflection and renewal allows for shifts in understanding: of ourselves, of the group and of the wider society (Trees and Kellas 2009). These shifts happen, in part, through relational learning and the challenging of assumptions (about ourselves and others, and our relationship to society) (Lykes and Scheib 2015).

Building from these story processes, further deconstruction of the power relations within different life stories builds an understanding of the structural injustices that affect people's shared realities. This approach to storytelling is undertaken to surface and question power relations, but through a creative, multimodal and inductive approach rather than one that is primarily analytical and relies on pre-established categories and classifications. The explicit attention to the relationships between the personal and the structural helps to avoid a potential weakness of personal storytelling, articulated by Sujatha Fernandes (2017). She argues that personal storytelling is often instrumentalized within a neoliberal discourse that erases issues of class and other structural causes of exclusion because it insulates wider hegemonic narratives from critique (Fernandes 2017). In this process, there is explicit attention to the relationship between the personal and structural dimensions of the story at different points.

Beyond the personal dimension, this approach to storytelling is also fundamentally relational. As Benjamin (2016) and Jackson (2013: 16) both argue, the telling of stories requires 'the presence of attentive others', or those who listen to the storytelling. This approach also implies considering more critically how we, as researchers, are positioned in relation to the topic we are researching and the people involved. Therefore, the process of storytelling shapes social relations and subjectivities: how people see one another and themselves, and what they expect as a consequence. In the context of Delft, storytelling also shaped the expectations of the relationships surrounding the research process, including between participants themselves, researchers, the NGO involved and the state (see Wheeler 2018). Therefore, storytelling used in this way can provide insight into how people claim a position and what expectations exist in relation to different actors. The next section turns to the storytelling process of the Delft Safety Group.

Personal storytelling, values formation and claiming a place

On a Monday in early March 2016, the twelve members of the Delft Safety Group arrived at a large warehouse space with blank white walls and plastic garden furniture. Over five days, I facilitated the group through a personal storytelling process[5]: the walls filled with drawings, instructions and ideas, and tables were covered in cuttings of paper, pieces of clay and scattered art materials. We started playfully with games and exercises that explore creativity, and open-ended maps of life experiences. Then we turned to the organising question for the storytelling: Tell a story from your own experience of when you felt secure or insecure in Delft. From this simple but grounded prompt, members of the group began to tell their stories. On that Monday afternoon, we sat in a story circle to listen to the first telling of each person's stories. Despite the many fears and possibilities for mistrust, especially given my position as a foreign white woman, the group took the opportunity to speak. This was more difficult for some than for others. Faquir,[6] one of the older men in the group, eventually became so overwhelmed by his own story about how his children were targeted by gang members that he stumbled into silence. We sat in silence and eventually he continued, speaking slowly and quietly. Next, Soeraya[7] reluctantly spoke about how her child was violated. As she continued to speak, more and more words began to stream out of her, and she spoke without stopping for over thirty minutes. After three hours, nearly every person in the room had become overwhelmed by the emotions of telling their stories.

When the first day of the workshop ended, the facilitation team sat with the trauma counsellor who had been observing the workshop to discuss how people were coping with the process and the situations in their lives. As we turned our attention to each person and how best to proceed, I noticed that my phone was constantly buzzing. I checked it and discovered that a child from my children's school had been raped and murdered at the park near our house, where I often go with my children to play – an unsettling reflection of the same stories I had listened to from Delft for the whole day. That evening, as I returned home to my middle-class neighbourhood, the people from Delft returned to their homes where murders (like that of the young woman in my neighbourhood) are a regular occurrence. I felt completely overwhelmed by the contradiction between my own sense of privilege and my emotional response to the stories: because of the stories emerging from the group from Delft, I felt a sense of loss of safety in my own life and for my family and the realisation of how difficult it is to live with violence. Over the rest of the week, the group from Delft worked through their personal stories through a range of creative techniques, and all twelve people came back every day and finished telling and creating their stories. I had a new sense of respect for how hard this was for them, as I struggled to deal with the emotions surrounding the death of my children's classmate.

As this vignette shows, the storytelling process is characterised by silences as much as by words, and by affect as much as events. The stories go through many versions, with different silences and things left unsaid, before they arrive at a final version. Yet through this process, storytellers are able to articulate their experiences without leaving out affect. And this form of storytelling allows the teller to claim recognition as a subject; to have their experiences recognised. Roland Barthes describes the shift of claiming subjectivity: 'I myself am my own symbol, I am the story which happens to me' (Barthes 1977: 56). The next section is an excerpt of the transcript of the final version of one of the stories that was told through this process.

Three broken hearts (A personal story told by soeraya davids)

When I got home that night after work, my youngest son told me that while they were out, one of our neighbour's children had lured him away and did something to him that was very sore. I could see the hurt and confusion on his face. He did not understand what had happened to him.

> A visit to the hospital a few hours later confirmed penetration. My 6-year-old son had been sexually molested. All anger came to me. I thought of the impact that this was going to make on my family. I tasted salt in my mouth and I realized I was crying in silence....
>
> The boy that hurt my son was with his parents and I could see the fear in his eyes when the officer arrested him and put him into the cells. He was 18. The next day, the judge at Bellville court said that there was no evidence to proceed. When I heard, I felt furious and heartbroken and I wanted to shout from where I was sitting. At that moment I wanted to leave Delft, but my life was there.
>
> My youngest son went for counselling at ChildLine. I never thought what happened would affect the older boy. Mostly I blamed myself for leaving them alone ... within a few months he had turned to drugs ... Maybe that was his only way of coping.
>
> Years went by when it really hit me that I never dealt with the pain of my son being molested. We always avoided the boy who did it.
>
> I wanted justice so that something good can come out of what happened to my son. But the truth is that the justice system failed me, and now we just have to carry on.

This version of Soeraya's story, which is the version she has chosen to share publicly, is compelling. But there is much that she left out of this version, that she spoke and conveyed in other moments of the storytelling process over several months. In the first version of this story that she recorded, the story had a different ending in which her neighbour went to jail for molesting her son. After a couple of weeks, she asked to re-record the ending because she said that she needed to accept the truth, which is that there was never any justice from the state for what happened. The case was thrown out due to a lack of evidence as a result of a procedural error in the clinic. For her, to even say those words and record them meant accepting that failure, and this was an extremely difficult step.

It is important to recognise that stories should not be taken as fixed or conclusive, but rather as constantly in play: as the perspective of the storyteller changes, so may their story. What is clear from the text of this version of her story is the role that the trauma associated with violence played in her life and that of her family, as well as the importance to her of the failure of the state to deliver justice. This is an example of how everyday experiences, including trauma, inform the demarcation of different values, which can surface through the storytelling process. Soeraya's story points to her expectation that the state should have been able to provide justice for what was done to her son, as well as her pain and disappointment that she was not able to rely on shared values to protect her son from her neighbour. While she had experienced the events in her story profoundly, their significance became clearer as she told and re-told the story, leading her to focus on her ending about the failure of justice. This is an example of how values emerge and become legible through the storytelling process, and also how the understanding of values and the story itself can change. The next section will explore in greater detail how this happens, and how these values can shed light on the understandings of justice through the example of trauma.

Collective narratives and collective claims

A central assumption of the social contract is that there are commonly held values that inform the responses of citizens to perceived failures of the state. Yet to understand and read the contestations and silences in the current political moment, greater attention needs to be paid to how, if at all, these shared understandings are constituted. Moving from examples of individual stories told through collective processes, this section explores how the values that emerged shape the expectations and positioning between citizens and the state through particular understandings of justice. Through attention to power and positionality, the storytelling process shifts from the articulation of experiences and affect and a claim for subjecthood; to making claims about what is expected from the state and from the research process itself. Thus, one can read the understandings of justice that emerge.

How does this play out in the practice of storytelling? Consider this example from the storytelling process with the group from Delft. After a version of the personal stories was concluded over a process of several months, we moved on to collectively analyse power within and across all the stories. After discussing different forms of power and performing short skits to illustrate the forms of power from examples in their own stories, the group began work on individual posters of power in their stories. The group cut out icons printed on paper to represent different forms of power. Using these icons and coloured yarn to represent power increases or decreases, each participant built a visual map of how power shifted in their own stories in relation to key characters and events.

The group spent hours making these posters, taking apart their own stories and thinking through what power means in their stories. Some members of the group struggled with the definitions of different types of power, and we spent time talking about the best word for translating the meanings into isiXhosa and Afrikaans. As individuals in the group went deeper into the analysis of their own story, they became more confident about how to define power in relation to specific characters in their own story. Each person worked through gluing the yarn at particular points, connecting the characters in their story to levels and forms of power, and explaining why they chose that point. There was an important moment of articulation of a new collective narrative when all the posters were displayed together. We tacked them up onto a wall so that we could sit together and look at them. As a group, we studied the posters with the colourful lines moving in many directions. One participant commented: 'Look at how often the storyteller felt powerless and how much power the police and the gangsters have in the stories'. From this moment, we moved into a new discussion of why power shifted in their stories, why they face the problems they described, who is responsible and what needs to change in South Africa and in Delft for other shifts to be possible. As the researcher, I deliberately moved the process into discussing the boundaries of state and personal responsibility. One participant, an older man, commented on the power analysis of the stories:

> In the power analysis of their stories, people chose 'power to', but in most stories they are still isolated with no help or direction. It shows that people are still wanting to do something, but the whole system that people are dealing with doesn't make it easy. People have tried different things, but it's hard to get a real solution to the situation.
>
> (Delft Safety Group member, during power analysis workshop)

Many disagreements emerged about the source of inequalities and what should be done about them – reflecting different political positions. Over subsequent months, the group developed a script for a collective film that emerged from their own analysis of their stories and the structural questions they prompted. The process of creating a collective story was framed by me, as the researcher, as being an opportunity for claim-making in relation to authorities. The group determined the nature of the claims and which specific authorities to address through their film. What emerged was how the experiences explored through personal stories are imbricated in structural dynamics. This was reflected within the group in terms of the reading of the personal experience of the stories through a collective lens of power. The development of a collective narrative by the group was also connected to the articulation of justice: a shared understanding about what is fair and just and the obligations of state institutions to uphold this.

This is an abbreviated script for one of these films:

Gangsters in uniform: A collective narrative by the Delft Safety Group[8]

This is the reality in Delft. We have gangsters and we have police. A lot of times we can't tell the difference. It's happening daily. It's becoming the norm. Police take bribes from gangsters and dealers. Our police, instead of protecting us from the gangsters, they are acting like gangsters. Innocent people are targeted and brutalised ... No one from the government or service providers helps you to report what happens ... our community service providers are not available when we need them. We put our complaint forms underneath the door as if we are submitting assignments ... Everywhere we go, there are closed doors, so we hold community meetings to inform people. We try to get community participation, to have our voices heard. Everyone comes to these meetings, even gangsters and corrupt police. We don't know where to turn because no one is helping us. The government is failing us. We have tried to work with the police through the neighbourhood watch, the CPF [Community Policing Forum], and every department you can think of but it's not working. Something needs to change in Delft police station, because people in Delft are dying and nothing is being done. We need police to be more accountable to the community. ... Community trust in police must be rebuilt ... We are looking for solutions for the problems we are facing on a daily basis in Delft.

As we prepared for a series of public events and discussions in which *Gangsters in Uniform* and Soeraya's story would be screened, I thought about that Monday in March when we sat in a story circle and I heard the first articulations of their stories. After a year of working together intensively, there were stronger relationships of trust between the members of the group and me, and multiple versions of their story in play. Some members of the group were gone: a young woman was in prison, a young man went back to the gang, an older man re-joined a corrupt faction of the police and a young woman was killed as she walked down the street by a speeding police car. There had been arguments and disagreements, but the members of the group still involved had become very committed to speaking publicly about the situation in Delft and what should be done to address it. They were motivated by a shared sense of justice: what they expected from each other, from the state and from the research process itself. The next section will explore the boundaries of this sense of justice through the specific example of trauma.

Trauma, values and the boundaries of justice

Trauma is very present in post-colonial contexts, but its consequences for democracy are not well understood. By looking more closely at how the experience of trauma is articulated through storytelling, it is possible to make more legible shared values and their implications for understandings of justice. In *The Empire of Trauma*, Fassin and Rechtman (2009: 2) set out two definitions of trauma, including the more narrow sense in the mental health field (of 'the traces left in the psyche') and the more widespread, popular meaning of trauma ('an open wound in the collective memory'). Thus, the construction of trauma also involves drawing lines of moral demarcation: it re-invents good or bad victims and creates a ranking of legitimacy among victims (Fassin and Rechtman 2009). Therefore, 'while trauma emerges in the context of an ethos of compassion that is characteristic of our era, it is also a tool used in a demand for justice' (Ibid.: 279). What emerged in this research process was the interplay between both forms of trauma.

Returning initially to the narrower definition of trauma, experiences of exclusion are often characterised by 'unsayable' emotions and affect, as described in the story circle with the Delft Safety Group. The process of storytelling involves ordering experiences and assigning them meaning through a narrative structure. Storytelling is a way that people 'make sense of their experiences and develop a sense of control and understanding' (Trees and Kellas 2009: 91). That is, most research on this topic within the field of psychology demonstrates how telling a story about 'difficult experiences' is valuable for physical health and psychological wellbeing (Monk et al. 1997; Pennebaker 1997; Pennebaker and Chung 2007). Specifically, telling stories about difficult experiences through a group process allows people to create organised sequences, identify emotions and label these experiences (Trees and Kellas 2009). The articulation of everyday stories of personal experience is crafted through 'layering' within the storytelling process, which allows for iteration between different forms of expression/creativity and relationships in order to create these sequences (Wheeler and Shahrokh 2018).

As demonstrated in her story, Soeraya faces an almost overwhelming weight of trauma. Her life has also been characterised by a lack of opportunity to address this trauma, despite working in roles that included counselling others. As described in the earlier vignette, there are many examples from across the Delft Safety Group of different forms of trauma. Trauma, in psychological terms, is related to dissociation, or not linking experience to thinking. Until this dissociation barrier is crossed, these experiences cannot be known (Hartman 2019). Therefore, the prevalence of trauma within contexts like Delft has significant political implications. In relation to shared understandings of justice, this means that until these experiences can become known, they cannot be part of the register of shared values, emotions and sentiments that ultimately underpin the social contract. As seen in Soeraya's story, the telling of stories in which trauma is present allows the experience to be assigned a name, and to take on its own value for the storyteller and those listening to the story. In Soeraya's story, she is able to articulate how the rape of her child was an injustice for her and her family and what this has meant to her. She is also able to name the failure of the state to deliver justice for what happened. In doing this, according to her, she was able to speak about what happened for the first time, and to articulate her own narrative around what this means for her; to claim her position as a citizen with a legitimate grievance who should be recognised by the state.

While trauma may be processed through storytelling, as described in the Delft Safety Group, it also presents challenges in terms of the ways in which the presentation of trauma can become a type of performance that implies a particular version of justice, as in the second interpretation of trauma set out earlier. Chris

Colvin traces an example of this process through 'traumatic storytelling', where the focus is on telling stories of victimhood during the anti-apartheid struggle with the expectation of articulating relations of obligation for redress in relation to current injustices (Colvin 2018). The Trauma Centre, a well-known organisation based in Cape Town, was established over twenty years ago to provide psychosocial services including counselling, especially for survivors of social crime, political violence and torture (Colvin 2018). It was founded during the transition to democracy to provide psychological support for those who experienced violence related to the anti-apartheid struggle, and has expanded since then to cover the effects of violence in South Africa more broadly. Within the context of the Trauma Centre and the history of anti-apartheid struggles in South Africa, as well as the current economic and political moment of sharpening inequality, Colvin (2018) argues that storytelling was undertaken in a way that ritualised moral economy – not only as a narrowly defined therapeutic process as was intended and framed by the Trauma Centre, but as an intentional act of articulating social relations and expectations in response to victimhood.

The moral economy of traumatic storytelling thus led to clashes between the Trauma Centre and the participants in their programmes, eventually leading to the formation of a separate social movement as the Trauma Centre did not respond to the claims for justice made by the storytellers in the ways they expected. Colvin (2018) describes how the Trauma Centre staff approached storytelling in a bounded way in relation to the therapeutic work necessary for dealing with trauma. This led to conflicts as the participants had expectations that went beyond the psychological work of healing, and the repeated work of telling stories of trauma became a practice that was not leading anywhere. Instead, telling their stories became a way of articulating their position of what they expected. This same impulse emerged with the Delft Safety Group over the course of the research process, and especially after the process of analysing their own stories, in which we deliberately connected this analysis to questions of why certain structures of inequality exist, for example through asking questions like: Why is there so much violence in Delft? Who is responsible for making Delft safe, and how can they be held accountable?

As Colvin's (2018) analysis of the Trauma Centre and the example from Delft show, there is a need to take care in analysing how narratives of trauma and victimhood are constructed through storytelling. The recognition of trauma and the claiming of victimhood can be understood as a claim for justice. While trauma 'offers a language in which to speak of the wounds of the past – of slavery, colonization or apartheid', trauma, when claimed by the protagonists themselves, 'becomes once again an argument in struggles for recognition of the plurality of memory – even if this violates historical reality' (Fassin and Rechtman 2009: 281).

In the public version of her story, Soeraya does not make a claim for justice – she names the failure of the state to provide justice. Instead, her story can be seen as a claim of her position as a political subject: in telling her story, she is demanding to be seen as a person who has experienced violence and who deserves to have justice. However, the articulation of her experiences of trauma through her personal story was part of what informed the collective narrative developed by the Delft Safety Group, in which specific claims are being made on the state. This progression, from personal to collective storytelling in the research process, reflected a different treatment of trauma. The first is about surfacing and claiming a position; the second, in the collective process, is about articulating what forms of justice are required in relation to that trauma.

These examples demonstrate how storytelling can make legible shared understandings of justice in terms of expectations of reciprocity, and also how storytelling constitutes specific understandings of social relations. Jackson (2013: 15) conceptualises this relationship as follows:

> When one tells stories, therefore, one is never simply giving voice to what is on one's own mind or in one's own interests; one is realizing, or objectifying, one's own experience in ways that others can relate to through experiences of their own. Stories are like the coins of the realm, the currency we implicitly agree to make the means of exchange, and, as such, a means of creating a viable social world.

Therefore, by analysing stories and storytelling, we gain access to a different register for understanding how political values are formed. What makes the values legible is the process of storytelling, in which there is an embedded 'means of exchange' around people's experiences. As this section has shown, there are many challenges to how this unfolds in contexts of extreme exclusion. In particular, trauma sets the boundaries of what can be known, told and recognised, and becomes a form of moral judgement (Fassin and Rechtman 2009).

Conclusion

While a clear case can be made for the importance of a storyteller narrating their experiences in terms of claiming and naming their own narrative, what is the relationship between these deeply personal stories and collective narratives? What is the political weight of these stories, when considered in relation to each other? What is their significance for questions of justice in the post-colonial context? While assuming that values are known and legible is problematic, this chapter explored the ways that values can be surfaced in a context characterised by multiple intersecting inequalities, and the implications of this process. A focus on storytelling can create space for embodied and complex experiences to become known and recognised. In the process, what has emerged is that values become known in layers, mirroring the methodological approach. The first layer is around how storytelling can lead to claims to subjecthood: to be seen by the other, and to have experiences and affect articulated and recognised. This is in some ways a subtle shift, and one that is less the focus of conventional research on politics and justice. Yet, this chapter has shown that by looking at how small stories are told and re-told, political values can be known. The example of Soeraya's story shows how multiple versions of the story can co-exist, and how her sense of what was just shifted in relation to these different versions. Thus, this chapter shows how the claiming of the story of the self is important for instantiating political values by bringing together events, affect and experiences through narrative. As these values are known, citizens themselves come into view more fully.

The second layer that emerged through this analysis focuses on the implications of these experiences for shared values and expectations for justice, implying a notion of the social contract. In the second layer, the emphasis shifts from the difficult task of articulating experiences, especially traumatic ones, towards making sense of these experiences in political terms. This movement between personal and collective is important for democratic politics in the post-colonial context because it allows a stronger connection between lived experience and wider political processes. That is, the 'spaces of everyday life and experiences of collective struggle can nurture alternative projects, forms of storytelling, and subjectivities' (Fernandes 2017: 168).[9] This chapter has shown how the spaces of everyday life can lend themselves to an alternative project in the form of new collective claims for justice, as with the Delft Safety Group's claims for justice in terms of insecurity. This chapter has traced the politics involved in moving from personal stories of the everyday to collective narratives and claims for justice. Ultimately, these collective narratives can illuminate emergent shared values and claims for justice, grounded in particular experiences.

Notes

1 See Price (Chapter 5, this volume) for a discussion of racial classification during apartheid.
2 This research process deliberately sought to cross historical racial categories.
3 Community Policing Forums (CPF) are constitutionally mandated spaces intended as mechanisms for citizen participation to increase accountability in policing. Many CPFs are not functional or are non-existent, and some have been captured by armed groups. There are also ongoing attempts to bring CPFs to life and improve their functioning – such as the 2016 White Paper on Safety and Security and the South African Civilian Secretariat for Police.
4 A collective account of this research can be found in Black et al. (2016).
5 Together with a group of researchers from Sustainable Livelihoods Foundation, including Nabeel Petersen, Rory Liedeman and Gill Black.
6 Name has been changed to protect anonymity.
7 Soeraya is a woman in her 40s who speaks Afrikaans and has lived in Delft for more than 15 years. She is an active member of the Neighbourhood Watch and the Community Policing Forum in Delft. She worked with the Victim Empowerment Room at the Delft police station until she was forced out because she spoke out publicly about corruption in the police. She is recognised by others as a leader within Delft. Despite these public roles, when she told this story, she had never spoken about the experiences before. She has given her consent for this story (and her role in the process) to be publicly available, using her real name.
8 The full film can be watched at: http://www.vimeo.com/192466247.
9 Fernandes (2017) notes this potential while cautioning against the potential for reverting to hegemonic agendas through storytelling – attention to the wider dynamics of power is important, particularly in relation to the post-colonial context.

References

Anciano, F., and L. Piper. 2019. *Democracy Disconnected: Participation and Governance in a City of the South*. London: Routledge.

Anderson, B. 2006. *Imagined Communities: Reflections on the Origin and Spread of Nationalism*. New York: Verso Books.

Bamberg, M., and A. Georgakopoulou. 2008. 'Small Stories as a New Perspective in Narrative and Identity Analysis', *Text & Talk – An Interdisciplinary Journal of Language, Discourse and Communication Studies* 28(3): 377.

Barthes, R. 1977. *Roland Barthes by Roland Barthes*. New York: Hill and Wang.

Benjamin, W. 2016. *The Storyteller: Tales Out of Loneliness*. London: Verso.

Black, G., N. Derakhshani, R. Liedeman, J. Wheeler and Members of the Delft Safety Group. 2016. *What We Live with Every Day Isn't Right: Partnerships for Greater Accountability and Safer Cities in South Africa*. Cape Town: Sustainable Livelihoods Foundation.

Boucher, D., and P. Kelly. 1994. *The Social Contract from Hobbes to Rawls*. London: Routledge.

Burns, D., et al. 2013. *Work with Us: How People and Organisations Can Catalyse Sustainable Change*. Brighton: IDS.

Collins, P., and S. Bilge. 2016. *Intersectionality*. Cambridge: Polity.

Colvin, C.J. 2018. *Traumatic Storytelling and Memory in Post-Apartheid South Africa: Performing Signs of Injury*. Oxon: Routledge.

Emberley, J.V. 2014. *The Testimonial Uncanny: Indigenous Storytelling, Knowledge and Reparative Practices*. Albany: SUNY Press.

Fassin, D., and R. Rechtman. 2009. *The Empire of Trauma: An Inquiry into the Condition of Victimhood*. Princeton and Oxford: Princeton University Press.

Fernandes, S. 2017. *Curated Stories: The Uses and Misuses of Storytelling*. Oxford: Oxford University Press.

Forster, E.M. 1927[1955]. *Aspects of the Novel*. San Diego, CA: Harvest.

García, O., and W. Li. 2014. 'Translanguaging', in C.A. Chapelle (ed.), *The Encyclopaedia of Applied Linguistics*. New Jersey: John Wiley and Sons, pp. 1–7.

Hartman, S. 2019. 'Hashtag Mania or Misadventures in the #Ultrapsychic', *Studies in Gender and Sexuality* 20(2): 84–100.

Jackson, M. 2013. *The Politics of Storytelling: Variations on a Theme by Hannah Arendt*. Copenhagen: Museum Tusculanum Press.

Kabeer, N. 2016. '"Leaving No One Behind": The Challenge of Intersecting Inequalities', in ISSC and IDS (eds.), *World Social Science Report 2016: Challenging Inequalities: Pathways to a Just World*. Paris: UNESCO Publishing.

Lykes, M.B., and H. Scheib. 2015. 'The Artistry of Emancipatory Practice: Photovoice, Creative Techniques, and Feminist Anti-Racist Participatory Action Research', in H. Bradbury (ed.), *SAGE Handbook of Action Research*. London: SAGE, pp. 131–142.

Monk, G., et al. 1997. *Narrative Therapy in Practice: The Archaeology of Hope*. San Francisco, CA: Jossey-Bass.

Nyamnjoh, F.B. 2012. '"Potted Plants in Greenhouses": A Critical Reflection on the Resilience of Colonial Education in Africa', *Journal of Asian and African Studies* 47(2): 129–154.

Pennebaker, J. 1997. 'Writing about Emotional Experiences as a Therapeutic Process', *Psychological Science* 8(3): 162–166.

Pennebaker, J., and C. Chung. 2007. 'Expressive Writing, Emotional Upheavals and Health', in A. Baum, T.A. Revenson and J.E. Singer (eds.), *Foundations of Health Psychology*. New York: Oxford University Press, pp. 263–284.

Piper, L., and J. Wheeler. 2016. 'Pervasive, But Not Politicised: Everyday Violence, Local Rule and Party Popularity in a Township in Cape Town', *South African Crime Quarterly* 55: 31–40.

Reavey, P. 2011. *Visual Methods in Psychology: Using and Interpreting Images in Qualitative Research*. London: Psychology Press.

Wheeler, J., and T. Shahrokh. 2018. '*The Role of Creativity and Storytelling to Address Power Inequalities and Structural Violence*'. Paper presented at the *International Sociological Association Conference*, Toronto, Canada.

Sonn, C.C., G. Stevens and N. Duncan. 2013. 'Decolonisation, Critical Methodologies and Why Stories Matter', in G. Stevens, N. Duncan and D. Hook (eds.), *Race, Memory and the Apartheid Archive*. London: Palgrave Macmillan, pp. 295–314.

Trees, A., and J.K. Kellas. 2009. 'Telling Tales: Enacting Family Relationships in Joint Storytelling about Difficult Family Experiences', *Western Journal of Communication* 73(1): 91–111.

Wacquant, L. 2008. *Urban Outcasts: A Comparative Sociology of Advanced Marginality*. Cambridge: Polity.

Wheeler, J. 2018. 'Troubling Transformation: Storytelling and Political Subjectivities in Cape Town, South Africa', *Critical African Studies* 10(3): 329–344.

Wheeler, J., and F. Bivens. 2021. 'Storytelling as Participatory Research' in D. Burns and J. Howard (eds.), *SAGE Handbook of Participatory Action Research*. London: SAGE.

Wheeler, J., T. Shahrokh and N. Derakhshani. 2018. 'Transformative Storywork: Creative Pathways for Transformation', in J. Servaes (ed.), *Handbook of Communication for Social Change*. Beijing: Springer, pp. 1–22.

Zepeda, S.J. 2014. 'Queer Xicana Indígena Cultural Production: Remembering through Oral and Visual Storytelling', *Decolonization: Indigeneity, Education and Society* 3(1): 119–141.

2 Silent citizens and resistant texts

Reading hidden narratives

Nobukhosi Ngwenya and Bettina von Lieres

Introduction

In this chapter, we pay attention to the ways in which marginalized groups are silenced by those in power, but also how their political values are represented in silent citizenship practices and unheard narratives. In South Africa, the problem of participation by poor and marginalized groups in politics is primarily one of silencing by contending social groups and political authorities with divergent conceptions of citizenship and the social contract. We identify three forms of silencing. The first account describes how long-standing livelihood practices become rendered illegal under neoliberal policies and are thus criminalized and deemed illegitimate. An understanding of the problem of silent citizenship must involve understanding the ways in which powerful social groups may silence certain forms of political agency by framing them as criminal. This framing may actually serve to shift the blame for marginality away from state policy choice to the marginalized themselves. This places the onus on the marginalized to drive change, but from a position in which whatever they say is already undermined. We use the example of illegal fishers' struggles for rights to illustrate criminalization practices. The second account makes a case for a conception of silencing through de-mobilization. Emergent due to incessant everyday experiences of non-recognition, coercion, repression, material deprivation and indignity; some citizens are silent not because they feel that the system does not work for them, but for a more fundamental reason: they do not see themselves as citizens, perhaps not even as any kind of social actor. The third account describes a conception of silencing through universalism. We argue that political values are often represented in social practices and narratives animated by these experiences of silencing, many of which remain unrecognized by mainstream, colonial and heteronormative Western epistemological and ontological perspectives and their dominant understandings of resistance.

We argue that our inability to hear and recognize long-standing discourses and practices of resistance by silenced groups – who are often rendered ethnic, traditional, marginalized or 'other' under the universal and modern episteme brought by colonialism – is due to the grounding of our research methods in colonial (read Eurocentric) epistemologies and ontologies, which shape all aspects of life (Mignolo 2018). This grounding renders us blind and deaf to resistant texts, understood in this chapter as expressions and narratives of everyday experiences of citizenship and power. Resistant texts tell stories that help us understand how different individuals and collectives in their everyday lives conceptualize political values such as justice and fairness. These stories also help us to recognize micro forms of resistance, usually ignored because they are indirect and subtle. Examples of resistant texts discussed in this chapter include traditional dress, beadwork, and land and building occupations by marginalized communities practising a politics of "quiet encroachment" (Bayat 2000).

The first section of this chapter outlines our understanding of silent citizens. We show how marginalized communities can be silenced in multiple ways: through criminalization, de-mobilization and universalism. We highlight that citizens are often silenced when their actions are examined through a colonial and heteronormative episteme, which often de-legitimizes the 'other' by criminalizing, de-mobilizing and universalizing. In the second section, we show how political values and conceptions of resistance manifest themselves in narratives and practices – such as resistant texts, which historically have been hidden and invisible. These include beadwork, traditional attire and social practices, such as illegal land or building occupation and the renaming of occupation sites. In the third section, we argue that recognizing and reading these practices and narratives allows us to understand how resistant texts represent political values. These texts, tools and practices reflect a rich diversity of political values and political subjectivities. These hidden values spur individuals and groups to act in the face of injustice.

Silent citizens and political values

In recent years, there has been a renewed focus on the connections between citizenship and silence by scholars in established democracies. The interest in silent citizens has emerged in response to the growing crisis of democracy, the weakening of civil society (Mercer 2002; Heller 2009) and the deepening of inequality globally. There are two broad schools of thought with regards to citizens' silence in established democracies. The first understands silence as a form of intentional and apathetic disengagement from democracy, whilst the second understands silent citizenship as a strategic political choice or a form of strategic disengagement. The first identifies 'the global phenomenon of a rapidly expanding underclass of silent citizens who are unaware of public issues, lack knowledge about public affairs, do not debate, deliberate, protest, or hold office and, most fundamentally, do not exercise their voice in elections' (Gest and Gray 2015: 465). Thus, this school, the larger of the two, understands silence as a form of cognitive disengagement from democracy. The second, less prominent school frames silent citizenship not so much in terms of cognitive disengagement, but as a strategic political choice – or strategic disengagement.

Gest and Gray (2015: 465–6) provide three examples of this. First, many citizens choose not to vote because they have no meaningful representative who reflects their views. Second, new forms of political engagement may replace voting as the perception grows that elected representatives account to their funders before their voters. Third, silence, especially in the form of not voting, might reflect the rise of more oppositional views and politics amongst the youth and minorities. All in all, then, the lack of engagement is a deliberate choice rather than due to apathy or despair. Notably, both schools frame the participation of poor and marginalized citizens in politics in terms of the political agency of these groups. Hence, when these groups do not participate, they are portrayed as silent. Underlying this assumption is the idea of liberal citizenship, in which all members of the state are rights-bearing individuals who are free to vote in elections, join political parties, stand for office, engage in debate in the media and organize and mobilize in civil society (Gest and Gray 2015). Under these conditions, the lack of action is imagined as either consent or some form of intentional silence through alienation and apathy or strategic choice to disengage. In this literature, then, the focus is largely on how citizens intentionally disengage, and not so much on the ways that states and other political authorities silence citizens from above.

Common across the literature on subaltern or popular politics in the Global South, by contrast, is the notion that political struggles by poor and marginalized groups confront resistance from both public authorities and contending social groups. This conflict is often intense and violent, and political engagement carries great risk to the bodies and minds of subaltern groups and especially their leaders. With some notable exceptions, the conditions for conflict remain high across the Global South as rural livelihoods decline and urbanization continues, often across national borders. At the same time, urbanization is not matched by growing employment in the cities. Unable to cope with the demand, urban systems of housing, electricity, water, sanitation and the like inevitably exclude large numbers of the poor, leading to significant levels of informal settlement, livelihoods and other practices. As Chatterjee (2004) notes, in such conditions of precarity and demand for resources, conflict with the state, by whose rules the urban poor cannot fully live, is inevitable. Further, it is also common with middle-class and other social groupings, including other formations of the poor and marginalized.

In this context of endemic conflict over basic urban needs, the notion of intentional silent citizenship is inadequate for understanding subaltern political engagement. This is because this body of literature tries to explain the political engagement of citizens primarily in terms of the purposeful political intent of these groups. Thus, the mainstream silent citizenship literature focuses on the apathy or strategic disengagement of poor or marginal groups. This approach assumes that these dispositions explain subsequent political actions or the lack of them. However, we suggest that sometimes politics comes first, and that poor and marginalized groups may not be silent due to apathy or strategic disengagement, but rather silenced by those in power in post-colonial contexts, like South Africa for example, through practices such as criminalization, de-mobilization and universalization.

Silent citizenship through criminalization

Although people may still speak and act, their words and actions are portrayed as illegitimate, and therefore beyond full inclusion in political consideration. This politics we term 'silencing through criminalization'. The construction of long-standing livelihood practices as now illegal effectively allows powerful groups to transfer blame for marginalization away from the policy choice of the state to the victims of the policy change. A great example of this politics is made by Anciano (Chapter 4, this volume), who outlines the criminalization of fishing practices in Hangberg, a long-standing fishing community of several thousand people in Hout Bay, Cape Town, South Africa.

Despite living off the sea for generations, the fishers of Hangberg increasingly found their livelihoods threatened by policy changes by the national government after 1994 (Anciano and Piper 2019). These policies allocated most of the fishing quotas and licences to large national and international fishing companies rather than to local communities. In pursuing policies informed by neoliberal and environmental concerns, rather than sustainable livelihood practices, the government unintentionally rewarded often white-owned big business to the detriment of black fishing communities like Hangberg.

Over the last twenty years, the fishers of Hangberg have exhausted every formal and informal channel to engage the state to change its policy – from elections, to lobbying and public protest, but with limited success (Ibid.: 127–31). While there has been some progress, with the South African government recognizing that fishing communities need a better deal, real change in policy has yet to materialize. Consequently, many of the fishers of Hangberg have chosen to continue to fish illegally or turn to drug or Abalone[1] smuggling to survive. Notably, many of the fishers of Hangberg refer to poaching as 'protest fishing'. The construction of long-standing livelihood practices as illegal effectively allows powerful groups to transfer blame for marginalisation away from the policy choice of the state to the victims of the policy change. Thus, fishers are transformed into criminals; the legitimacy of their political agency is undermined, albeit not entirely eradicated.

Informed by experiences in the Global South, Chatterjee (2004) famously described this ambiguous relationship to the state from a position of informality and semi-illegality as typical of the condition of the urban poor across the Global South. It helps illustrate how legitimate struggles by citizens can be taken out of the domain of legitimate politics and into hidden practices or non-public, more silent practices. Central to challenging de-politicization as criminalization is for the marginalized to disrupt the symbolic order through renaming who they are and what they want. For the community of Hangberg, this involves renaming poachers as protest fishers and insisting on more just policy regimes around fishing allocations and licences.

Silent citizenship through demobilization

Citizens can also be silenced when their experiences of non-recognition, marginalization and violence by public authorities and/or contending social groups is simply entrenched and prolonged. Under such conditions and over time, marginalized communities can disconnect from practices and expectations of rights and public political protest. Examples include communities with high levels of social fragmentation and inaction because of prolonged state inattention and community violence.

Karl van Holdt (2012) argues that community violence in South Africa has deeply corrosive effects that lead to a wider demobilization of large numbers of citizens, in addition to wider social fragmentation and ongoing cycles of violence. Runciman (2015: 962) has shown that the demobilization of new social movements has been a significant feature of subaltern politics in post-apartheid South Africa. She argues that a key characteristic of contemporary protest is that it has occurred largely outside of the organizational forms that would conventionally be termed social movement organization. As a result, collective demands for service delivery are often fragmented and disconnected from one another. The highly localized and isolated nature and the sporadic and un-coordinated trajectories of the protests has not received much attention. Claims for a wider rebellion of the poor or a wider insurrectionary citizenship movement fail to acknowledge the fragile trajectories of these protests. Many communities experience long periods of demobilization and depoliticization because of state and community violence, and misrecognition. There has been relatively little research on how state inattention and violence demobilize, depoliticize and silence communities.

Silent citizenship through universalism

This leads us to the third form of silencing – silencing through universalism. This form of silent citizenship is about the inability of researchers to hear the long-standing discourses of resistance by groups rendered ethnic or traditional under the universal and modern episteme brought by colonialism. Citizens are not heard because their practices and narratives cannot be heard from a colonial and heteronormative standpoint. Citizens or communities are silenced through epistemic non-recognition. The post-colonial context of South Africa is one of demographic, linguistic and epistemic diversity. Nevertheless, colonial forms of knowledge dominate through the institutionalization of political and economic systems imposed under colonialism and inspired by heteronormative standards. Citizens are rendered silent or invisible when their actions are examined exclusively through a colonial and heteronormative episteme. This lens blinds and deafens us to what Sommer (1992) has referred to as 'resistant texts' and results in what we have termed silencing through universalism. We now turn to a set of examples of how political values are formed and represented in moments of unseen resistance and unheard narratives. Sometimes these hidden practices and values are not fixed or easily known, as they might be embedded in everyday practices that are not easily discernible.

Resistant texts, hidden practices and political values

Resistant texts are 'everyday expressions of endogeneity' (Winkler 2018: 589). That is, they are stories and narratives of endogenous[2] (read African in this chapter) experiences and perspectives. As Kessi and Boonzaier (2017) argue, resistant texts are tools with which power and systems of oppression can be understood, resisted and transformed. Resistant texts are based on and illustrate alternative and often unheard ways of knowing, being and acting.

Resistant texts often tell stories and offer descriptive narrations of past events in a person's life. Shenhav (2015: 3) notes that:

> Stories help us to define who we are, who we are not, where we want to go, and where we came from. Stories contain repertoires of reactions to social situations. Accordingly, they can structure people's conceptions of society or serve as models for societal and personal behaviour. In these and many other ways, stories can guide people's actions. Stories are also vital for remembering, but they are no less important for disremembering or forgetting.

Narratives and stories allow us to make sense of facts, 'identify their significance, and even, when they are painful or unpleasant, to accept them and live with them' (Gabriel 2004: 62). A story is a narration of events. Therefore, a narrative is one representation of a story. To construct a narrative, one needs to select events to speak about and decide how these events relate to each other. A change in event order results in the creation of a new narrative on the same story. The formation of narratives requires that one draw on different types of knowledge – from general knowledge to knowledge on how people interact in (certain) social situations and memories, as well as knowledge of the listener's expectations (Hudson and Shapiro 1991). The narrator must then weave these different types of knowledge into a narrative. Therefore, social narratives are open. They do not have a definitive beginning, middle and end unlike narratives, in the dramaturgical sense. In fusing different knowledge(s), the narrator infuses facts with meaning (Gabriel 2004). It is these facts and meanings that we, as the audience, listen for (see Wheeler, Chapter 1, this volume). It is also this process of infusing facts with meaning that makes social narratives, particularly resistant texts, that much harder to read by those who are not familiar with these types of texts.

Resistant texts are stories that organize hope and despair (Sandercock 2003) and help to mobilize individuals or communities to act collectively. By reading resistant texts, we can begin to discover the political subjectivities of millions of people across the Global South through the stories and narratives they tell about their lives and experiences. Resistant texts can help us find answers to our 'why' questions at all levels – experiential, intersubjective, relational and representational – in which resistant texts operate. However, as we listen and by virtue of our academic training, we tend to identify and search (probe) for the missing facts. Second, we – both the listener and the narrator (Shenhav 2015) – look for clues regarding the significance of the missing facts. Third, we tend to silence not only the facts, but also stories we consider insignificant (Gabriel 2004) and, central to the argument put forward in this chapter, we also unintentionally silence meanings attributed to particular experiences that do not conform with our ways of knowing (epistemologies), ontology (being) and acting (axiology).[3] These meanings are often found in resistant texts, which reveal how acts of resistance and protest are structured in ways that are unfamiliar to the eye that is trained to examine the world using positivist research methods.

Resistance through transgressions of dress conventions

Magwaza (2001), for example, writes of the use of traditional dress by a young Zulu woman to share her heartbreak and anger as her boyfriend celebrated his marriage to another woman. To express herself, the young woman wore *isidwaba* (a black animal skin skirt) and *udikwe* or *isibhodiya* (a red apron), both normally worn by married women. This visual or metaphorical form of expression (Biyela 2013) is used, as Berglund (1976: 18) notes, to 'voice thoughts, experiences and concepts, and to do so intelligibly', sometimes through a transgression of dress conventions – as a means of personal and social protest. The message conveyed through the traditional attire can be, and often is, amplified through the addition of symbolic elements, such as beads. *iZangoma* (diviners), for example, wear beaded pieces, which help identify them and their work. The beads also signify the spirits (*abaphansi*) and the importance of the *Sangoma*'s relationship with the ancestors. A novice *Sangoma* often wears a single string of white beads around his/her elbows, ankles, wrists and head (Katsande 2014). As they become more experienced, the intricacy of the beadwork they adorn increases. Furthermore, *iZangoma* can create beadwork for clients who need support in a particular, often spiritual, area of their life. The colours of each beadwork piece differ, subsequently, depending on the ailment and the individual's characteristics (Katsande 2014).

Protesting through beadwork is possible only because the beaded elements of traditional regalia continue to be used by all genders to convey symbolic messages and tell (their) stories, histories and myths in a visual and descriptive manner during social interactions (Zungu 2000; Magwaza 2001). The triangle is the simplest shape used in Zulu beadwork to denote gender and material status. Each corner of the triangle signifies the father, the mother and the child. By carefully patterning the colours and symbols, messages are written in the first person which, in conjunction with the spoken word:

> may produce magic actions or events once it is uttered with authority … The perceived bead colours in Zulu may summarise a situation, pass a judgement, or offer a course of action within a particular context. For example, a message sent in [*sic*] a white-beaded string can be a consolation in difficulties and a guide when choice is made and can also express a morality suited to the human behaviour in a Zulu cultural context.
>
> (Biyela 2015: 469–70)

The combination of 'art and linguistic logic' defines society and simultaneously recognizes some people but not others. It also helps to frame issues publicly through a system of symbolic order (Zulu Beadwork Language 2015). Such public and visual framing of issues through systems of symbolic order is unfamiliar to observers interpreting the symbols from a colonial epistemological standpoint. Yet, colonial epistemology continues to be used to categorize the lived experiences of individuals and/or communities across the globe (Spivak 1988; Todd 2016; Mignolo 2018), even though it does not enable us to understand fully resistant texts such as those discussed earlier. The consequence of unfamiliarity is that it often translates into non-recognition and, inter alia, silencing. This, in turn, leads to actors whose actions are informed by non-hegemonic epistemologies being labelled erroneously as silent citizens. Despite voicing their grievances in a visual manner, actors are silenced because their actions do not conform to constructed representations of participation and protest (Kiguwa and Ally 2018). This silencing may be unintentional; nevertheless, it is a real exercise of power.

Beadwork is a visual form of communication; a method of communication which, through the transgression of dress conventions, not only narrates an individual's history and social position but also is a form of social commentary and protest. However, it is an act of resistance that does not conform to hegemonic conceptualizations of protest within the colonial episteme, which portends we are all party to a single, universal social contract. The examination of resistance activities through a colonial gaze, underpinned by the additional assumption of universal, rights-based citizenship in post-colonial states – even nominal democracies like South Africa – not only silences citizens but also masks the existence of social contracts between individuals. The argument in this section, therefore, illustrates that social contracts exist between communities and the state, as well as between individuals. The rupture of social contracts between individuals also elicit expressions of disapproval and/or objection from the aggrieved parties.

Resistance through (Re)naming

Resistant texts also reveal how social justice and, consequently, political subjectivity are conceived in multiple and sometimes unfamiliar ways between particular individuals or groups at the local level. This is not to say that there are no deep differences in interests, perspectives and agendas – what Watson (2003: 403) refers to as 'conflicting rationalities' – within groups (read communities) often thought to be homogenous (see West 2005; Pickerill 2009). Our interest, however, is on the differences *between* groups, namely the state and occupiers who have organized themselves as Reclaim the City (RTC). Reclaim the City is a Cape Town-based social movement of tenants and workers who have launched several campaigns to stop their displacement from Cape Town's inner city areas due to gentrification, and to access affordable housing in the inner city. Brought together by their experiences of homelessness/landlessness, indignity, marginalization and inequality, these individuals have mobilized and occupied vacant buildings in various parts of the inner city to address their housing needs and to counter narratives positioning land and building occupiers as bad citizens. Reclaim the City has, in its words, 'reclaimed' (read occupied) three government-owned houses in the inner city – the central business district, Green Point and Woodstock – to meet members' desperate needs for housing.

As Huchzermeyer (2003) notes, occupations remain a key driver of development in South African cities. They are also the key mechanism through which poorer urban residents access housing. Despite acceptance of this observation, the experiences of homelessness/landlessness, marginalization and indignity – which lead individuals or groups to occupy – remain largely invisible. This is due to skewed power relations, which give political elites the ability to craft a single story; a story in which occupiers' claims are defined as illegal and, consequently, illegitimate if they are not made through the policy-ascribed channels for restitution and redistribution.

In addition to casting occupations as illegal, narratives constructed by political elites serve the dual purpose of characterizing occupiers as bad citizens. Bad citizens are those who, tired of waiting or having realized that the state will not be able to meet their needs in their lifetime, have taken it upon themselves to meet their housing needs on their own by claiming land that is seemingly underutilized or vacant. The result of bad citizens' actions are informal settlements and occupied buildings, which are more common in Johannesburg (Dugard and Ngwenya 2019). However, and given that a significant number of people live in informal settlements, therein lies an interesting paradox in the characterization of the good citizen as one who conducts themselves within the ambits of the law. As Oldfield and Greyling (2015) note, the politics of waiting often require subversion in the interim in order to survive, rendering them both legitimate and unethical citizens at the same time. Yet, the state continues to pit good citizens against bad citizens.

Occupiers are pitted against good citizens who practice a politics of waiting, which entails placing one's name on the housing database, waiting and making do until it is your turn to receive a state subsidized house (Oldfield and Greyling 2015). In so doing, and through various narratives, the government marks certain individuals' needs – that is, the needs of good citizens – as being more legitimate than the needs of bad citizens through the criminalization of informality (Rao 2010). Thus, the state creates a narrative of good beneficiaries and bad beneficiaries whilst simultaneously constructing narratives around who belongs in the city (and the country), and consequently who bears rights – even though the Constitution unequivocally holds that all South Africans hold a series of inalienable rights.

To counter these narratives of bad citizenship, and to assert belonging, occupiers rename the buildings or land they have reclaimed. For example, one occupation site whose residents have subsequently been evicted by the City of Cape Town's Anti-Land Invasion Unit in conjunction with other law enforcement agencies, has been renamed *Izwe Lethu*, which means 'our nation'. According to the City of Cape Town's records, the site was formally established as Graceland. The RTC occupiers have renamed the Woodstock Hospital and Helen Bowden Nurse's Home after prominent figures in the struggle for liberation, namely Cissie Gool House and Ahmed Kathrada House, respectively. The names reflect, first, the occupiers' commitment to securing their rights. As Charol Jacobs (cited in McCool 2017: online), who is part of the Woodstock Hospital occupation, states, 'My granny lived in District Six all her life … She fought for her rights, and my mum after her fought for her rights, so I'm not going to leave my rights behind'.

The renaming of the occupation sites is, as Paulo Freire in 1987 (cited in Tuhiwai-Smith 1999: 81) noted, akin to 'claiming the world and claiming those ways of viewing the world that count as legitimate'. Hence, in renaming the occupation sites, occupiers are also speaking of their individual and collective experiences of injustice whilst highlighting that they have not lost hope. The names also reflect the expressed desire for belonging and a 'better future in South Africa' (Occupier (2017), interviewed by Ngwenya, 22 September 2017).

Un- and relearning in order to read resistant texts and silent resistances

One of the first steps to learning how to read resistant texts is unlearning: unlearning how we have been taught to view the world, and unlearning the need to know how to read all texts (Winkler 2018). Resistant texts require us to deepen our search for meaning and explanations beyond the scope of our personal and academic knowledge. Thus, unlearning, like belonging, is a process (Yuval-Davis 2006). It is 'always work in progress to be constantly enriched with new encounters and new relationships, and never to be confined by geography or boundaries, political or disciplinary' (Olukoshi and Nyamnjoh 2011: 5). A process that involves acknowledging endogenous knowledge systems still informs how millions of people know and move through the world (Olukoshi and Nyamnjoh 2011). The term endogenous is preferred over the term indigenous, as the former encapsulates the 'dynamism, negotiability, adaptability and capacity for autonomy and interdependence, creativity and innovation in African societies and beyond' (Nyamnjoh 2012: 136). This definition, when understood in relation to systems of knowledge production, consequently puts forward the idea of African[4] knowledges as dynamic (Winkler 2018).

The term 'endogenous' also allows one to acknowledge that different bodies of knowledge influence each other (Winkler 2018). As Nyamnjoh (2012: 148–9) argues, 'Epistemological conviviality and interconnection are possible precisely in light of the spirit of tolerance for which Africans are renowned, and in recognition that there are no final answers to perplexing questions in a dynamic world'. Notwithstanding, we need to be careful not to succumb to the tendency to characterize society in terms of the traditional-modern dualism. Dualism is not a characteristic of the popular epistemologies, which, according to Nyamnjoh (2012), are used to attribute meaning to their experiences and various concepts. Popular epistemologies:

> create room for why questions, and for 'magical interpretations' where there are no obvious explanations to material predicaments (Moore and Sanders 2001). In them, reality is more than meets the eye. It is larger than logic. Far from subscribing to rigid dichotomies, popular epistemologies build bridges between the so-called natural and supernatural, physical and metaphysical, rational and irrational, objective and subjective, scientific and superstitious, nature and culture, visible and invisible, real and unreal, explainable and inexplicable. Inherent in the approaches is the recognition of the impossibility for anything to be one without also being the other.
>
> (Nyamnjoh 2012: 131)

As we unlearn, we gain the ability to embrace popular epistemologies. In doing so, we are able to more fully understand and distinguish between various resistant texts. We are also able to '[pay] attention to the often contradictory values that guide and sustain livelihood practices, through which cultural domination is reproduced or altered' (Palomera and Vetta 2016: 413). These contradictory values, which are narrated in resistant texts, can be unearthed using storytelling as a methodology.

Storytelling can be utilized to uncover the stories of how and why the event(s) happened. Storytelling can also be utilized, as is already the case in the planning profession, to tell what Throgmorton (2003: 127) termed 'future-oriented stories'. Storytelling, whether it is understood as a process or as the use of story to facilitate process (Sandercock 2003), is often thought of as 'a woman's way of knowing, as inferior, lacking in rigor' (Sandercock 2003: 12). In response to this criticism, Sandercock (2003) argues that story plays a pivotal role in the process of making invisible experiences legible. Resistant texts render the invisible visible by countering mainstream narratives and focusing on actors who often are not acknowledged when the official version of history is narrated (Sandercock 1998).

Stories are everywhere, and with the advent of social media, many more people are able to share their stories, and to (re)present themselves. Herein lies the power of stories, narratives and, ultimately, storytelling. Through stories, we are able to discover the teller's political subjectivities (Jackson 2002; Wheeler, Chapter 1, this volume). We are also able to discover how reactions to injustices are increasingly dictated by moral values; values that are based on African and colonial (read Western) ways of being and moving through the world. These values, although not directly observable, can be gleaned through the stories individuals and/or groups tell about themselves and their experiences. Put differently, stories reveal the meaning of human existence, the human psyche, and our understanding of the storyteller's individual and collective purpose (Brooker 2004).

Yet, most of us (even those of us educated in tertiary institutions in the Global South) are trained, almost exclusively, to uphold positivist standards of enquiry, and thus, we are unable to read resistant texts.[5] This is assuming that we are able to recognize them in the first instance. As Nyamnjoh (2012: 131) notes:

> The science (natural and social) inspired by such an epistemology has tended to celebrate dichotomies, dualisms, teleologies and analogies, dismissing anything that does not make sense in Cartesian and behaviourist terms, confining to religion and metaphysics what it cannot explain and disqualifying as non-scientific more inclusive epistemologies. This epistemology's logic is simple and problematic: it sacrifices pluriversity for university and imposes a one best way of attaining singular and universal truth.

This Cartesian rationalism continues to be used to understand and categorize the lived experiences of individuals and/or communities across the globe. It consequently blinds us to resistant texts by fixing definitions of social constructs. Once defined, social constructs have a huge influence on the meanings attributed to the stories we read. They also greatly influence which stories, independent of historical context, we seek out (selection bias).

As a number of scholars have pointed out, the failure to recognize historical differences gives rise to universal theories that 'fail to account for the diversity of global experiences, and particularly experiences and conditions of the Global South' (Lawhorn et al. 2016: 1612). This is problematic in a number of contexts, such as our own, in which there is no clean cut – so to speak – between the spiritual and secular. *IZangoma*, for example, also play a number of social and political roles in their communities – meaning their roles are not limited to divination. Here, we have an example of the blurring of the line between the physical world and the spiritual that colonial epistemologies typically do not recognize. This shortcoming serves to depoliticize and silence the 'other' in a manner that not only erases other ways of being from liberal conceptualizations of resistance, but also masks the existence of multiple and contending forms of social contracts between the state and individuals, as well as between individuals. This is exemplified in our discussion of protest fishing, land occupation and traditional beadwork, respectively. The assumption that we are party to a universal social contract, ultimately, can have the effect of limiting citizens' access to certain liberties or result in the erasure of individuals. As Césare argued in 1956 (cited in Hountondji 1973[1996]: 30), it is possible to lose oneself through 'dilution in the "universal"'.

Conclusion

Diverse silencing of poor and marginalized groups, and the co-existence of contending social contracts by various groups and the state, call into question sanguine assumptions of liberal democratic citizenship in post-colonial conditions. The reality, it seems, is more about the engagement between various groups and political authorities who invoke various and often contending 'social contracts', many of which silence the poor and marginalized. Yet, in telling resistant texts, those who have experienced injustice are assigning meaning to their experiences. In so doing, protestors co-construct their political subjectivities. Collins (1986) notes that this subject positioning can be a hindrance and/or a benefit to the individual occupying that status. However, this can be uncovered only through the stories that this individual tells. The structure of the text and the events related in the story, as well as the language used to tell the story, reveal significant clues about the worldviews people cohabit (Nyamnjoh 2012).

We must, therefore, unlearn and (re)learn to read resistant texts, as these texts narrate alternative histories, values and definitions of justice which are indicative of different epistemologies and ontologies. By unlearning how we read texts, we are able to identify the different worldviews people cohabit, even if the 'culture-specific, understandings of the order of things and the human condition' (Olukoshi and Nyamnjoh 2011: 10) make us uncomfortable. This discomfort stems from not knowing (how to read) a text, be it in the form of protest fishing, a beaded necklace, traditional attire or the occupation of a building.

Notes

1 Abalone, also known as perlemoen (*Haliotis midae*), is a pale coloured form of a large sea snail, housed in a flat, shiny shell. It is traded live, frozen, tinned, canned or dried.

2 The term 'endogenous' is used in lieu of the term 'indigenous' in this chapter as it enables the representation of African cultures as fluid and characterized by

'dynamism, negotiability, adaptability and capacity for autonomy and interdependence, creativity and innovation in African societies and beyond' (Nyamnjoh 2012: 136).

3 This act of silencing happens on both the part of the storyteller and the listener. Shenhav (2015: 1) points out that: 'Narratives usually "magnetize" their audiences to their agendas, making them dwell on specific events and occurrences'.

4 For a summary of the debates around what we mean by 'African', see Higgs (2010).

5 We use the term 'text' rather loosely in this chapter to refer to the content of stories and narratives, irrespective of physical form. We do this purposely, recognising that in some cultures across the continent, knowledge is passed from generation to generation orally.

References

Anciano, F., and L. Piper. 2019. 'Poaching the Bay', in F. Anciano and L. Piper (eds.), *Democracy Disconnected: Participation and Governance in a City of the South*. London: Routledge, pp. 122–142.

Bayat, A. 2000. 'From "Dangerous Classes" to "Quiet Rebela": The Politics of the Urban Subaltern in the Global South', *International Sociology* 15(3): 533–557.

Berglund, A. 1976. *Zulu Thought-Patterns and Symbolism*. Uppsala: Swedish Institute of Missionary Research.

Biyela, N.G. 2013. 'Colour Metaphor in Zulu Culture: Courtship Communication in Beads', *Journal of Sociology and Social Anthropology* 3(10): 37–41.

Biyela, N.G. 2015. 'Women and the Art of Speaking with the Beads: The Traditional "Zulu Love Letter"', *Journal of Sociology and Social Anthropology* 6(4): 469–473.

Brooker, C. 2004. *The Seven Basic Plots: Why We Tell Stories*. London: Continuum.

Chatterjee, P. 2004. *The Politics of the Governed: Reflections on Popular Politics in Most of the World*. New York: Columbia University Press.

Collins, P.H. 1986. 'Learning from the Outsider Within: The Sociological Significance of Black Feminist Thought', *Social Problems* 33(6): 18–35.

Dugard, J., and M. Ngwenya. 2019. 'Property in a Time of Transition: An Examination of Perceptions, Navigations and Constructions of Property Relations among Unlawful Occupiers in Johannesburg's Inner City', *Urban Studies* 56(6): 1165–1181.

Gabriel, Y. 2004. 'Narratives, Stories, Texts', in D. Grant, C. Hardy, C. Oswick and L. Putnam (eds.), *The SAGE Handbook of Organisational Discourse*. London: SAGE Publications Ltd, pp. 61–79.

Gest, J., and S.W.D. Gray. 2015. 'Silent Citizenship: The Politics of Marginality in Unequal Democracies', *Citizenship Studies* 19(5): 465–473.

Heller, P. 2009. 'Democratic Deepening in India and South Africa', *Journal of Asian and African Studies* 44(1): 123–149.

Higgs, P. 2010. 'Towards an Indigenous African Epistemology of Community in Education Research', *Procedia Social and Behavioural Sciences* 2: 2414–2421.

Hountondji, P. 1973[1996]). *African Philosophy: Myth and Reality*. Bloomington: Indiana University Press.

Huchzermeyer, M. 2003. 'A Legacy of Control? The Capital Subsidy for Housing, and Informal Settlement Intervention in South Africa', *International Journal of Urban and Regional Research* 27(3): 591–612.

Hudson, J.A., and L.R. Shapiro. 1991. 'From Knowing to Telling: The Development of Children's Scripts, Stories and Personal Narratives', in A. McCabe and C. Peterson (eds.), *Developing Narrative Structure*. New Jersey: Lawrence Erlbaum Associates, pp. 89–136.

Jackson, M. 2002. *The Politics of Storytelling: Violence, Transgression and Intersubjectivity*. Copenhagen: Museum Tusculanum Press.

Katsande, R. 2014. 'The Importance of Beads and Beadwork in African Culture'. *Wilderness Safaris*. Accessed January 20 2019. https://wilderness-safaris.com/blog/posts/the-importance-of-beads-and-beadwork-in-african-culture/.

Kessi, S., and F. Boonzaier. 2017. 'Resistance and Transformation in Postcolonial Contexts', in C. Howarth and E. Andreouli (eds.), *The Social Psychology of Everyday Politics*. London and New York: Routledge, pp. 116–130.

Kiguwa, P., and Y. Ally. 2018. 'Constructed Representations of Street Protest Violence: Speaking Violence, Speaking Race', *Peace and Conflict: Journal of Peace Psychology* 24(1): 36–43.

Lawhorn, M., et al. 2016. 'Unlearning (Un)Located Ideas in the Provincialisation of Urban Theory', *Regional Studies* 50(9): 1611–1622.

Magwaza, T. 2001. 'Private Transgressions: The Visual Voice of Zulu Women', *Agenda: Empowering Women for Gender Equity* 49: 25–32.

McCool, A. 2017. '"End Spatial Apartheid": Why Housing Activists Are Occupying Cape Town'. *The Guardian*. Accessed June 12 2019. https://www.theguardian.com/cities/2017/may/25/spatial-apartheid-housing-activists-occupy-cape-town-gentrification.

Mercer, C. 2002. 'NGOs, Civil Society and Democratisation: A Critical Review of the Literature', *Progress in Development Studies* 2(1): 5–22.

Mignolo, W.D. 2018. 'Decoloniality and Phenomenology: The Geopolitics of Knowing and Epistemic/Ontological Colonial Differences', *Journal of Speculative Philosophy* 32(3): 360–387.

Moore, H.L., and T. Sanders. 2001. *Magical Interpretations, Material Realities: Modernity, Witchcraft and the Occult in Postcolonial Africa*. London: Routledge.

Nyamnjoh, F.B. 2012. '"Potted Plants in Greenhouses": A Critical Reflection on the Resilience of Colonial Education in Africa', *Journal of Asian and African Studies* 47(2): 129–154.

Oldfield, S., and S. Greyling. 2015. 'Waiting for the State: A Politics of Housing in South Africa', *Environment and Planning A: Economy and Space* 47(5): 1100–1112.

Olukoshi, A., and F. Nyamnjoh. 2011. 'The Postcolonial Turn: An Introduction', in R. Devisch and F. Nyamnjoh (eds.), *The Postcolonial Turn*. Netherlands: Langaa and African Studies Centre, pp. 1–27.

Palomera, J., and T. Vetta. 2016. 'Moral Economy: Rethinking a Radical Concept', *Anthropological Theory* 16(4): 413–432.

Pickerill, J. 2009. 'Finding Common Ground? Spaces of Dialogue and the Negotiation of Indigenous Interests in Environmental Campaigns in Australia', *Geoforum* 40: 66–79.

Rao, U. 2010. 'Making the Global City: Urban Citizenship at the Margins of Delhi', *Ethnos* 75(4): 402–424.

Runciman, C. 2015. 'The Decline of the Anti-Privatisation Forum in the Midst of South Africa's "Rebellion of the Poor"', *Current Sociology* 63(7): 961–979.

Sandercock, L. 1998. *Making the Invisible Visible*. Berkeley, CA: University of California Press.

Sandercock, L. 2003. 'Out of the Closet: The Importance of Stories and Storytelling in Planning Practice', *Planning Theory and Practice* 4(1): 11–28.

Shenhav, S.R. 2015. *Analysing Social Narratives*. New York: Routledge.

Sommer, D. 1992. 'Resistant Texts and Incompetent Readers', *Latin American Literary Review* 20(40): 104–108.

Spivak, G.C. 1988. 'Can the Subaltern Speak?' in C. Nelson and L. Grossberg (eds.), *Marxism and the Interpretation of Culture*. Basingstoke: Macmillan Education, pp. 271–313.

Throgmorton, J.A. 2003. 'Planning as Persuasive Storytelling in a Global-Scale Web of Relationships', *Planning Theory* 2(2): 125–151.

Todd, Z. 2016. 'An Indigenous Feminist's Take on the Ontological Turn: "Ontology" is Just Another Word for Colonialism', *Journal of Historical Sociology* 29(1): 4–22.

Tuhiwai-Smith, L. 1999. *Decolonising Methodologies: Research and Indigenous Peoples.* London: Zed Books Ltd.

Van Holdt, K. 2012. 'The Violence of Order, Orders of Violence: Between Fanon and Bourdieu', *Current Sociology* 61(2): 112–131.

Watson, V. 2003. 'Conflicting Rationalities: Implications for Planning Theory and Ethics', *Planning Theory and Practice* 4(4): 395–407.

West, P. 2005. 'Translation, Value, and Space: Theorising an Ethnographic and Engaged Environmental Anthropology', *American Anthropologist* 107: 632–642.

Winkler, T. 2018. 'Black Texts on White Paper: Learning to See Resistant Texts as an Approach towards Decolonising Planning', *Planning Theory* 17(4): 1–17.

Yuval-Davis, N. 2006. 'Belonging and the Politics of Belonging', *Patterns of Prejudice* 40(3): 197–214.

Zulu Beadwork Language. 2015. 'Beadwork in the Zulu Cultural Tradition'. *Eloquent Elegance*. Accessed December 3 2019. http://marques.co.za/clients/zulu/bead.htm.

Zungu, B.P.K. 2000. 'Meaning behind the Use and Wearing of Traditional Beadwork at Msinga Area', MA Dissertation. Durban: University of Durban-Westville.

3 A moral economy of citizenship in post-apartheid South Africa?

S.J. Cooper-Knock

Introduction

During the transition from apartheid, many jurists, activists and politicians sought to create a new legal order that would forge and protect a democratic South Africa. In a country historically defined by oppressive legal regimes, this was a monumental step forward. The new constitution that emerged from this transition included substantial socio-economic rights for citizens. This chapter explores the degree to which those constitutional rights represented a moral economy between citizens and the state in post-apartheid South Africa, and what forms of political relationships and resistance emerged from this particular constitutional settlement.

Since E.P. Thompson engaged with the idea of the moral economy back in 1971, use of the term has expanded exponentially. At its broadest, the term is close to a 'floating signifier' (Hall 1996), referring to any intersection between norms and economic activity, however loosely defined. Much creative and insightful academic work has come out of this broader field of study. For the purposes of this chapter, however, I will focus on work that has attempted to engage with Thompson's (1971) original use of the term. My interest lies primarily in the degree to which the concept of a 'moral economy' is useful for understanding politics in post-apartheid South Africa.

After sketching out Thompson's conceptualisation of the moral economy, I will argue that we need to refine his ideas to make them relevant to post-apartheid South Africa. First, I draw on the established literature surrounding Thompson's work in order to temper his ideas about 'the economy', moving away from any notion that battles over moral economies emerge only at moments of complete economic rupture. Second, I shift the temporal frame of Thompson's work. Thompson's moral economy is conservative: the crowds that Thompson studied were trying to preserve norms that had historically existed. In contrast, my focus is on the moral economy of citizenship that emerged in the transition from apartheid, grounded in constitutional law. This moral economy was rooted in a new vision for an inclusive, humane future intended to serve those who had been systematically oppressed and excluded under the apartheid state. It was a pre-figurative moral economy.

Having put these ideas in place, I explore the extent to which a moral economy lens helps us in analysing struggles over adequate housing in South Africa, in particular the struggles of informal settlement residents. I do so by exploring its relevance to understanding resistance and the acquisition of resources. In terms of resistance, I argue that a 'moral economy of citizenship' has some analytical purchase in helping us to understand the political environment in South Africa, as well as the role that the government has imagined for its citizens. In doing so, the concept usefully builds on work that explores the 'politics of patience' and the 'politics of waiting' in post-apartheid South Africa and beyond. Ultimately, however, the idea of a 'moral economy of citizenship' is limiting because it encourages us to focus on a singular logic of statehood and tempts us to overlook the normatively fickle nature of the law in post-apartheid South Africa. To gain a more comprehensive perspective of the relationship between law and people's lived experiences, we need to also explore the moral economies that people attempt to forge themselves in resistance to the law: the permissive spaces that people use to gain access to the resources they were promised. This chapter concludes by explaining this notion of 'permissive space' and the political dynamics it enables us to uncover.

Defining the moral economy

Thompson developed the notion of the moral economy as a means of understanding eighteenth-century riots in England. These were not, Thompson argued, 'simple responses to economic stimuli' (Thompson 1971: 76). Instead, they represented a 'highly-complex form of direct popular action' (Ibid.: 78), which opposed changes in the production and trade of bread. Previously, regulations had restricted this market in ways that protected the ability of poor people to feed themselves, and people resisted their relaxation, claiming such protections 'as their right and their heritage' (Ibid.: 88). This resistance was, periodically at least, recognised by 'paternalist' authorities who benefitted from the status quo and sought to both placate the poor and limit the rise of middlemen (Ibid.). For a time, whilst the economic tide was turning, the courts upheld symbolic cases against those thwarting market restrictions. Thus, the moral economy 'had an ideal existence and also a fragmentary real existence' (Ibid.: 88). In cases where legal authorities declined to act, the crowds took over, acting as effective price-setters for traders, millers and bakers.

Eventually, Thompson argues, the moral economy of the late eighteenth century was displaced by the 'new political economy' of Adam Smith and others, which was 'disinfested of intrusive moral imperatives' (Ibid.: 90). The rise of this 'demoralised economy' intersected with heightened fears of Jacobinism in the wake of the French Revolution (Ibid.). Consequently, by the nineteenth century, state authorities were unwilling to intervene on the issue and fiercely opposed any attempts by the crowd to do so (Ibid.). Economic faith in the unencumbered market had fused powerfully with political fears of popular intervention. Whereas crowds had been met with 'parley and concession' in the past, they now faced immediate 'repression' (Ibid.: 129). Moreover, what the poor might have seen as the maintenance of a socio-economic pact, the authorities saw as 'sedition' (Ibid.: 129).

As the summary above demonstrates, Thompson developed the notion of the 'moral economy' to elucidate a very particular set of struggles. Subsequent work has explored whether this original application withstands scrutiny and whether the concept is transferable across time and space. Of particular relevance to this chapter are two inter-related critiques that emerged from discussions attempting to expand and refine Thompson's work. I discuss these below and add a third critique, which is crucial if we are to adapt the moral economy framework for post-apartheid South Africa.

The first critique of Thompson's approach focuses on the false binary that is created between the 'demoralised economy' of the free market and the moral economy of its predecessor. Thompson did this, scholars suggest, by comparing the economic *ideology* of Adam Smith against the economic *reality* of the eighteenth-century marketplace. This was a mistake. When analysing moral economies, Götz highlights, we must not 'mistake postulates of current economic theory for actual modern practice' (Götz 2015: 158). If he had explored the economic realities of the nineteenth century instead, Thompson would have seen a *differently moralised* economy rather than a *demoralised* economy. Certainly, many exchanges were given over to the market, but moral economies continued to exist, albeit sometimes over different goods and on different terms.

Critics of Thompson have also questioned whether it is possible to speak of a wholesale economic transition between modern and pre-modern economics at all. Contra Karl Polanyi's work, we must appreciate that any definitive shift from an 'embedded' to an 'autonomous' economy obscures the reality that 'all economies (ancient and modern) are simultaneously economizing and normative in nature, although in varying degrees' (Arnold 2001: 88). That said, contestations over the moral economy are not limited to points of transition between 'pre-modern' and 'modern' or 'non-market' and 'market' societies – should these even exist in any bounded sense. Therefore, scholars of moral economies need not be limited to wholesale battles over market incorporation. Instead, they can usefully explore far more contained struggles. As Thomas Arnold argues, 'politically significant moral indignation' is as (if not more) likely to emerge within, rather than between, economies and societies (2001: 85).

Following Arnold's (2001) rationale, this chapter sees contestation over the moral economy as a battle over a particular good or group of goods, rather than a battle over an epochal economic shift. That does not, however, mean that all battles over goods can be framed in terms of the moral economy. In order to fall into the latter category, three criteria must be met. First, the good(s) in question must be claimed outside the market. Second, those claims should be supported by a wider population than the claimants alone (Thompson 1993: 338). As Thompson argues, the crowds in the nineteenth century 'were informed by the belief that they were defending traditional rights or customs; and, in general, that they were supported by the wider consensus of the community' (Thompson 1971: 78). Finally, the good(s) should carry particular significance (Götz 2015: 154). For Arnold (2001), a moral economy forms specifically over a social good. This need not be what Arnold calls an 'inherently social good' such as air (Ibid.: 93). By 'social good', Arnold simply refers to a good that sits 'at the intersection of nested sets of meaning and value'

(Ibid.: 85). A social good 'establishes and symbolises an important sense of self' (Ibid.: 90). Furthermore, the 'patterned use, display, and exchange of social goods', Arnold contends, 'place people in specific social, economic and even political relations' (Ibid.: 91). These 'identity orienting, role-shaping' goods hold a value that stretches 'beyond [their] purely material or commercial properties' (Ibid.: 91). In the context of South Africa, it could be argued that any of the basic services associated with basic needs and adequate housing (e.g. water, electricity, housing) fulfil this idea of a social good. These have a clear use value, but they are also central tenets of the dignified life promised to all citizens at the end of apartheid.

The two points above leave us with a refined, contained and more generally applicable idea of the 'moral economy'. In order for the term to be useful in the context of post-apartheid South Africa, however, I would argue that we need to add a final point. For Thompson (1971), the 'moral universe' from which people's claims emerged was fundamentally rooted in the past. Thus, eighteenth-century crowds drew on ideas of local customary rights over production and price protections that had been established centuries before. Likewise, in James Scott's (1977) exploration of the moral economy of peasants in Malaysia, norms are pre-established practices. Similarly, in her informative study of Johannesburg, Prishani Naidoo (2010: 135) states that debates over the moral economy are based in 'the assertion of particular traditions, customs, values, beliefs, and commitments held in common in the past' as well as current demands on the state. Acknowledging the importance of historical practices and ideas in shaping the parameters of a moral economy in South Africa is undoubtedly important. That said, the notion is not as straightforwardly conservative as it was for E.P. Thompson or James Scott. In both Malaysia and England, people sought to preserve an already-existing set of obligations (Thompson 1971; Scott 1977). In post-apartheid South Africa, however, the moral economies upon which people drew were often prefigurative rather than established, particularly as they related to the state.[1]

Making this temporal divergence explicit is analytically important because political struggles that seek to preserve what has already existed are fundamentally different from those that seek to realise what has not yet materialised. In South Africa, the moral economy of citizenship emerged from concepts of freedom and dignity that were forged in the midst of multiple struggles. These ideas started from the premise that people's humanity and dignity were being denied and imagined what it would mean to recognise that humanity and allow it to thrive (Biko 1978; Mandela 2007; Mandyoli 2012; Walker 1991). As we will see, the incoming ANC government in 1994 sought to inscribe something of these visions into the new constitution, which outlined the rights of citizens under the new dispensation. To that extent, they created a moral economy of citizenship and called citizens to adopt a 'politics of patience'.[2] In this context, the framework of a moral economy can tell us something useful about a particular dynamic between statehood and citizenship as cultivated by the ANC government. Ultimately, however, we need move beyond this concept to understand the complexities of statehood, citizenship and the law in post-apartheid South Africa.

Forging a constitutional moral economy in post-apartheid South Africa

Whilst lawlessness was certainly a feature of government oppression in apartheid South Africa, much of the brutality of the regime was carefully codified into the 'wicked system of law' prior to 1994 (Chaskalson 2003: 590). Apartheid was by no means original in this regard – preceding regimes had also made oppressive use of the law – but the National Party built on this legacy and expanded the 'deviance allowed by the rules' (Brogden and Nijhar 1998: 90). In the midst of this 'quintessentially unjust legal order' (Wacks 1984: 183), those who managed to use the legal system for just ends were vastly outnumbered by those who were oppressed by it.

In the wake of apartheid, many jurists, activists and politicians sought to reconcile their legal and moral universe. Their aim was to create a new legal order that would forge and protect a democratic South Africa. The freedom and equality that was to define this democratic dispensation had a strong material underpinning, which reflected the material concerns of the struggle against apartheid. As Kader Asmal, a CODESA delegate for the ANC, highlighted: 'The struggle for liberation in South Africa was not only a struggle for the right to vote, to move, to marry or to love. It has always been a struggle for freedom from hunger, poverty, landlessness, and homelessness' (Asmal, in Christiansen 2007: 3). Polling before the elections in 1994 echoed this verdict – people's demands for a dignified life drew on a range of key services from housing to healthcare, and from education to electricity (Barchiesi 2011: 70).

In keeping with this view, the constitutional order that emerged in the mid-1990s provided citizens with socio-economic rights. If moral economies exist when 'community membership supersedes price as a basis of entitlement [to a commodity]' (Tilly, in Thompson 1993: 338), then we can think of socio-economic rights as being the bounds of the moral economy of citizenship proposed by post-apartheid leaders. As such, the constitution became the means through which the ANC pledged to make good on the pre-figurative imaginaries of freedom and dignity forged in the anti-apartheid struggle. A moral economy demarcated by legal citizenship became the foundation of the 'Better Life for All' promised by the ANC during the first post-apartheid election campaign.

The remainder of this chapter explores the ways in which it might be fruitful to think of a moral economy in these terms. My discussions will focus primarily on the struggles over housing by informal settlement residents in eThekwini, South Africa, which I have followed over the last ten years. The right to 'adequate housing' is covered in Section 26 of the Constitution, and the government is ordered to move towards the 'progressive realisation' of this right, as its resources allow. The definition of 'adequate housing' is contested, but the courts have acknowledged that 'housing entails more than bricks and mortar', and under some interpretations this includes a right to electricity (RSA & Ors v Grootboom & Ors 2000; Dugard 2009: 266). This article will explore the degree to which thinking of a moral economy of citizenship in relationship to housing can help us understand politics, law and life in post-apartheid South Africa.

I argue that such an approach can help us to understand the power of the constitutional promise and what might be called a 'politics of patience'. The latter powerfully shaped the first decade of post-apartheid housing struggles in eThekwini and continues to influence how people explain their struggles today. Ultimately, however, the moral economy of citizenship is only one relational dynamic amongst many that exist in post-apartheid South Africa. It draws our attention to a singular relationship between statehood, citizenship and the law. In reality, the government views South Africans as consumers as much as it sees them as citizens. For their part, those who live within informal settlements find that the law is normatively fickle: it is never inherently just (Makhulu 2015). Therefore, engagement with the law necessarily becomes more strategic and citizens create their own everyday moral economies by negotiating 'permissive spaces' outside the law. Such tactics mean that performing patience neither fully encapsulates people's strategies to secure state services nor their relationships with the motley mix of institutions and actors that constitute 'the state' (Oldfield and Greyling 2015: 1110).

Political promises and the 'politics of patience'

Winning power in 1994, the ANC government sought to draw civil society leaders into the state and demobilise the oppositional political movements that had played such a powerful role in the final years of apartheid (Bond 2001; Desai 2002; Chance 2018). The time for the politics of protest, they argued, had passed: Citizens could now trust that their government was acting in their best interests. The constitution (finalised in 1996) was presented as proof of this commitment – evidence that the country was realising the promises of liberation grounded in socio-economic rights (such as the right to housing), which belonged to every South African citizen. This right to material goods through the constitution is what I refer to as the 'moral economy of citizenship'.[3] As stipulated in the constitution, such rights were to be realised over time and according to the resources of the government. Thus, what the country needed in the interim, officials claimed, was patient collaboration. To act otherwise was to work against the country's democratic future: a political and moral failing (Mbali 2013).

This is the dynamic upon which the 'politics of patience' rested: The constitution had declared that citizens *would* have access to goods in the future. In this context, South Africans did not need to actually possess the goods named in the constitution in order to believe that the moral economy of citizenship was functional. When a moral economy is conservative, people act to defend what already exists. When a moral economy is pre-figurative, people must reflect on what is yet to come. Thus, to say that a moral economy of citizenship existed in South Africa was to argue that the government was making all reasonable efforts to deliver the goods that they had promised. It follows that to say that the moral economy of citizenship was broken, was to say more about the character of government than the presence or absence of a particular good. By asking citizens to be patient, the ANC was asking them to maintain their faith in the moral economy of citizenship and, by extension, the government itself.

Before going further, it is important to note that the 'politics of patience' has meant different things to a variety of scholars, and to position this chapter amidst those competing definitions. Arjun Appadurai (2001) explores an alliance between several social movements in Mumbai, India, whose work intersects (in varied ways) with housing struggles in the city. These movements have attempted to forge a long-term relationship with one another that slowly nurtures grassroots knowledge and builds careful consensus on key areas between diverse organisations. He calls this work the 'politics of patience'. Patience, in this context, is seen as a commitment that opens up space for radical, people-centred politics that engages with parties and state departments tactically: a strategic means to a radical end.

In contrast, Javier Auyero (2012) explores the ways in which patience signifies the closure of radical space and the reproduction of state domination. His ethnographic research focuses on Argentina where, he argues, poor citizens are forced to wait for the assistance they need from the state (2012: 124). In the corridors of power, he contends, citizens realise that they must dutifully wait for a state bureaucracy that is inefficient, overloaded and capricious. The 'submissive set of dispositions' (Auyero 2012: 9) that they adopt to do so is often gendered because women are usually given the task of waiting (Auyero 2012: 126; Carswell et al. 2019). Whilst Auyero (2012: 158) briefly explores the possibility of resistance, his emphasis remains on the compliance that waiting generates. The Argentinian state, he argues, lowers people's expectations of immediate service whilst maintaining just enough hope to ensure their acquiescence (2012: 25, 68).

For the purposes of this chapter, I argue that attempts by the state to encourage a 'politics of patience' are attempts to frame patience as the work of the moral, responsible citizen. This framing is ultimately an attempt to keep citizens in their place and, in doing so, to reinforce state domination. Take, for example, the response of then ANC Secretary General Gwede Mantashe to residents of Tshepisong informal settlement, who complained about their living conditions. Mantashe claimed that the ANC had not had enough time to deliver houses to everyone and pointedly informed residents that 'patience is a virtue' (Gallens 2017). If, as Dobler (2020: 3) argues, we think of waiting as encompassing multiple forms of action and attitude, then we can think of 'patience' as a call by the state to a particular attitude towards waiting.

In this context, patience is inherently political. Far from being a passive act, patience is the political work done by citizens who continue to suffer amidst the stark material realities of their everyday life but nonetheless engage with the state on its own terms (Auyero 2012). This may be because they actively believe that the state has their best interests at heart, or it may be because they believe it is politically expedient to behave 'as if' this were the case (Wedeen 2003). In situations where citizens are behaving 'as if' they are patient, we can think of waiting as being simultaneously a site in which state dominance is reproduced (Auyero 2012) and resistance ferments (Dawson 2014: 869; Jeffrey 2008: 956; Lombard 2013: 818; Mujere 2019: 3).

In the early years of the post-apartheid regime, informal settlement residents engaged heavily in the 'politics of patience'. Whilst some resistance could be 'glimpsed' in the 1990s (Nthambeleni 2008),[4] it was arguably not until the 2000s that this resistance became meaningfully organised and deeply rooted in eThekwini's informal settlements, and beyond. Perhaps even more remarkable is the persistent importance that the idea of patience played in these mobilisations: even when people stopped engaging with the state on its own terms, they continued to stress the lengths to which had previously gone to do so. Members of the shack-dweller's movement Abahlali baseMjondolo, for example, consistently highlight how long they have waited for service delivery prior to protesting. The phrase 'we are tired of waiting' is a common refrain in community meetings.[5] In protests, press releases and meetings, members often stress exactly how long they have engaged in the 'politics of patience'. The Good Hope Settlement in Germiston, for example, introduced their plans for occupation by stating, 'The Good Hope Settlement in Germiston is now twenty-two years old. Some people have lived their whole lives in the settlement waiting for the promises of land and housing to be fulfilled'. Picking up on this trend, Nigel Gibson (2016) also stressed that in 2005, when the movement emerged, 'the shack dwellers had been waiting patiently for Nelson Mandela's historic 1994-election promise of housing to be realized'. These statements are not limited to Abahlali baseMjondolo. The Mandela Park Backyarders in Khayelitsha, for example, stated in 2010 that they were mobilising because they had not received the housing that they had been promised. 'When we finally spoke to our councillor', they explained, 'he told us to keep waiting. And we've been waiting and waiting, and we are tired of waiting. We are tired of his empty promises' (Mandela Park Backyarders 2010).

Analysing the moral economy of citizenship in South Africa enables us to realise why this is important: having faith in the promise of delivery was the political work assigned to citizens by the ANC during the transition. Citizens were told that a failure to practice the 'politics of patience' at that point was not just a political flaw, it was also a moral failing. It is telling that over two decades later, those who are protesting the emptiness of this moral economy still feel the need to prove that they did the political work of patience that was asked of them. The 'politics of patience' has, I would suggest, powerfully shaped the first decade of post-apartheid grassroots politics and continues to influence how people frame and legitimate their struggles for a more dignified life.

The moral economy meets the moral consumer

In the previous section, I have suggested that the 'politics of patience', which was strongly shaped by the ANC government's pursuit of a pre-figurative moral economy, has had a lasting effect on struggles over the legitimacy of resistance and protest in South Africa. As the remainder of this chapter argues, however, it highlights one relationship and repertoire of legitimacy amongst many. Ultimately, the danger of engaging with the idea of a moral economy of citizenship is that it encourages us to think in terms of a singular logic of statehood: the state as material provider of basic needs to its citizens, anchored in law. In contrast, anthropologies of statehood in recent years have pushed us to think about the multiple (and sometimes conflicting) logics that drive the discourse and actions of state representatives in everyday life (Joseph and Nugent 1994; Hansen and Stepputat 2001; Hagmann and Peclard 2010). These conflicting logics are not just the result of contestation between tiers, departments or factions of government. They are often held simultaneously by the same state institution or official (Gupta and Sharma 2006). Below, I highlight two logics of statehood that exist in tension in South Africa: the state as material provider to its citizens, and the state as protector and advocate of the market. As I will argue in the final section of this chapter, the law is an instrument of statehood in both cases. Both constrain the space in which people can autonomously claim some measure of the goods they were promised. Consequently, informal settlement residents may feel pushed to negotiate a 'permissive space' against the law to secure some of the resources that they have been promised within it. The moral economy of citizenship, in other words, is often realised (in whole or in part) beyond the state and, sometimes, in opposition to it.

In the following section, it is worth noting, we will focus on only two of the many visions that the ANC government has of South Africans[6]: the vision of the rights-bearing citizen, and that of the moral, responsible consumer. Within the constraints of this chapter, I must necessarily simplify these visions and overlook others.[7] Nonetheless, these two visions are telling. Since the transition, they have substantively shaped how the ANC has thought about service delivery and the access that people could or should be able to have to goods. These two idea(l)s have co-existed throughout the post-apartheid era. Thus, in the same year that Mandela spoke passionately about freedoms being grounded in legal rights, he also launched a government campaign that spoke of 'the freedom to pay for what we use' (Chance 2018: 54). Since 1994, payment for services has been framed as a central duty for citizens: without this payment, the government argued, services could not be provided for those in greatest need (Chance 2018: 55). Indeed, it was arguably during the first seven years of the post-apartheid period that the idea of payment for services was at its height, unmediated by indigent policies that were designed to secure basic services for the country's poorest citizens (Ruiters 2018: 171). Officials spoke of a 'culture of non-payment', which had emerged from the rent and service delivery boycotts at the end of apartheid (Bond 2001). Whilst such boycotts had played a valuable role in undermining an illegitimate government, they argued, non-payment in post-apartheid South Africa would only damage the realisation of democracy and the re-distribution of basic services.

This rhetoric has been sustained despite three clear limitations. First, by framing the issue as a 'culture of non-payment', officials were dismissing non-payment as the apolitical reflex of irresponsible citizens (Gumede 2008; Khunou 2002: 66). In reality, however, non-payment was not a reflex: it emerged from multiple motives. For some, such as the Soweto Electricity Crisis Committee (SECC),[8] non-payment stemmed from the belief that basic services should be free at the point of access (Nthambeleni 2008: 158). That social movement, forged in 2000, represented 'an alternative voice and logic to that of commodification … emerging out of the everyday struggles for survival of ordinary residents of Soweto' (Naidoo and Veriava 2009: 322).

Second, officials were presuming that people could pay for certain services. As research by Grace Khunou (2002) and others has demonstrated, this was not necessarily the case. To return to the issue of electricity, prices had risen in recent years against a backdrop of growing unemployment. Nefale (2004) found that electricity claimed a quarter of the income of households in Soweto. This difficulty has only increased over time. In the decade following the transition, for example, there was a price rise of around 131% for non-commercial consumers (Greenberg 2009: 101).

Third, their discourse suggested that existing rights-based delivery actually met the needs of the poorest in society. If we look at free basic electricity or free basic water provision, however, we see that this is actually far from sufficient for people's everyday lives. Free Basic Electricity provides eligible households with 50 kWh per month, which allows residents to fuel four 60-watt lights daily for seven hours per day. Meanwhile, Free Basic Water allows for a usage of 25 litres per individual per day. As Ruiters (2009: 269) claims, these measures 're-inscribe' rather than relieve marginalisation. Moreover, in many informal settlements, there are no formal electricity connections or water distribution systems to ensure that such allocations are secured. Remembering both the insufficiency of the Free Basic Allowance and the inaccessibility of authorised state infrastructure is crucial when we return to citizens' attempts to claim goods autonomously.

Despite these difficulties, the discourse of the 'moral consumer' remains powerful. Over the past two-and-a-half decades, it has been used to justify stark material situations which undermine the moral economy of citizenship. In positioning the moral economy of citizenship as being in conflict with the figure of the moral consumer, I diverge from Prishani Naidoo's (2010) conceptualisation of the moral economy in her valuable work on the construction of 'the poor' in Johannesburg. Naidoo argues that the figure of the moral consumer represents a 'refashioning' of the moral economy to include market principles (2010: 259). In this chapter, however, I understand the figure of the moral consumer as being incompatible with the concept of the moral economy because I understand the latter as a system that works outside or beyond the logic of market exchange. To incorporate market logics is, therefore, to crowd out rather than refashion the moral economy.[9]

To say that the moral economy and the moral consumer cannot coexist in the same singular exchange is not to say that these two logics do not coexist over time, space and services. As I argue above, both of these idea(l)s have been in play since the transition and it is vital that they are both kept within the analytical frame. In this sense, I agree with Gillian Hart's (2014) argument that South Africa is far from being a homogenous neoliberal state – particularly since the early 2000s – and we lose much of our understanding of statehood in South Africa if we suggest otherwise. Similarly, it is important that we do not frame protest as simply rejecting the 'moral consumer' in favour of the 'moral economy of citizenship'. Reality is more complex: the idea of the moral consumer has influenced the ways in which citizens have framed their protests over housing and dignity in South Africa. Whilst some informal settlement residents have defended informal connections to electricity or water as a right, other people's claims have included the figure of the consumer.

Take Busisiwe, for example, an informal settlement resident I interviewed[10] in a settlement where residents had created informal electricity connections. These connections were periodically destroyed by the Metro Police. Government campaigns accused informal settlement residents of stealing electricity. 'We are not stealing electricity', Busisiwe objected, 'we are taking electricity ... deep inside the people never see it as the municipality sees it – as theft – because we never refused to pay'. Here, Busisiwe is arguing that people within informal settlements in eThekwini have essentially been denied the opportunity to be consumers – rather than being denied their rights as citizens – because they have not been formally connected to the electricity grid. This, she suggests, leads her to reject the idea that what they are doing is illegal. Similarly, as Mnikelo Ndabankulu, then spokesperson for Abahlali baseMjondolo, explained, 'We do not call it an illegal connection, we call it a people's connection'. 'People have never said they cannot pay for electricity', he argued, 'it is the government who is unwilling to connect for them'.[11]

The law meets the moral economy

Reference to the policing of electricity connections above brings us to the focus of the final section of this chapter: law and the moral economy. On the one hand, what I have termed 'the moral economy of citizenship' is fundamentally grounded in law. In this sense, the law secures people's access to certain social goods. On the other hand, laws and regulations also limit people's access to these same goods. In situations where the state does not deliver, citizens are left with little legal space for autonomous action due to the laws that uphold market exchange (e.g. property rights) and the municipal bylaws that forbid unauthorised[12] connections to service infrastructure. Legal protection for those making informal connections and constructions (such as the legal limits on evictions) is limited.[13]

Many of those who autonomously seek to access some or all of the goods and services promised in the constitution often find themselves in violation of laws and bylaws. If anything, the criminalisation of autonomous service connections has increased in recent years. Take, for example, the emergence in 2015 of the Criminal Matters Amendment Act No. 18. This made it illegal to 'tamper with' or 'destroy' any 'essential infrastructure' or collude with others to do so. Those convicted are liable to 'a maximum of thirty years imprisonment or, in the case of a corporate body ... a fine not exceeding R100 million'. The attraction for many legislators is that this law can be used to target criminal syndicates involved in the large-scale theft of copper wire. Eskom and the South African Local Government Association, however, seem to want to use this legislation to tackle low-level tampering with electricity infrastructure by individual households. As one ANC representative objected in the public hearings, 'Eskom seems to be putting people from informal settlements and "izinyoka"[14] on the same level' (RSA 2015). In summary, the law creates a foundation for multiple logics of statehood and citizenship, access and ownership. Consequently, from the perspective of many informal settlement residents, the law is normatively fickle. Those who are tired of exercising a politics of patience may well look beyond the law to secure their basic needs (Oldfield and Greyling 2015: 1108).

To the extent that informal settlement residents do seek to secure basic services illegally, what does this mean for our understanding of the moral economy of citizenship? One could argue that, lacking affordable state services, informal settlements have secured their own informal realisation of that same moral economy, grounded in their constitutional rights. For this interpretation to hold, however, people would need to justify their access to goods on the basis of their unrealised constitutional rights. This, in turn, means two things. First, claimants would need to see themselves *as* citizens (even if they are not, in practice, officially recognised as such). Second, their claim to resources would need to be made within a rights-based frame. In practice, this is not always the case. Some members of Abahlali baseMjondolo, for example, think of resources such as land and water as God-given. Consequently, they see the marketisation and/or private ownership of these resources as 'a sin' (Motha, in Pithouse 2014: 141). Rights-based language does not, in other words, have a monopoly over people's justifications for claiming goods. Sometimes, rights-based language coexists alongside other logics of claim-making. At other times, they crowd them out altogether. Ultimately, whether or not a particular claim to socially significant goods falls within the moral economy of citizenship is an open question that can be determined only by the motivations and understandings of those involved in the act of claim making.

Furthermore, for there to be a moral economy in play, one party's claim to resources must be acknowledged by others (Thompson 1993; Arnold 2001; Götz 2015). The question then becomes, what counts as recognition? E.P. Thompson places much weight on court decisions and official documents as evidence of recognition. Undoubtedly this is due, in part, to the fact that historical studies of this era have to infer broader trends from limited sources. Nonetheless, Thompson's work allows us to consider a minimalist definition of recognition. For the purposes of this chapter, I argue that people's informal claim to a moral economy of citizenship is upheld when they are able to defend their access to goods or resources to fellow residents and state officials on the basis of their constitutional rights. Of course, just like the judges, politicians and other officials that Thompson explores, the individuals they encounter may have mixed motives for accepting people's claims to a particular good. Nonetheless, at some level, the legitimacy of people's claim to goods factors into an official's decision.[15]

Not all informal claims to goods are illegal. As explained earlier, however, informal settlement residents trying to autonomously secure resources like housing, water, electricity often find themselves in contravention of laws and bylaws. I call the space in which people negotiate illegal access without sanction from state or non-state actors, 'permissive space' (Cooper-Knock 2018). Within permissive space, illegal practices are allowed to continue unabated and may be conferred with legitimacy. The idea of permissive space is a flexible one. Those within permissive spaces can act as sovereigns, but this sovereignty is as fragile as the negotiated space itself. The negotiations that create permissive spaces range from amicable to antagonistic. Negotiation, in this context, simply emphasises that permissive spaces are not unilaterally imposed, nor are they immutable. Instead, they are established in an ongoing exchange. This exchange need not be symmetrical, and the 'consensus' reached in such negotiations may be fraught and shallow.

In some cases, the borderline between 'decision' and 'negotiation' may appear thin. However, to some degree, impunity is always unstable, open to the possibility of (re)negotiation at different times, scales and spaces. Permissive spaces can be relatively broad in the acts they allow and can be relatively stable over time. Conversely, they may also be narrow, fleeting and unstable. Much depends on the degree of permission granted and the power of those granting it: both of which are mutable. The breadth and stability of permissive space often shifts meaningfully over time. And the moral economies that they protect shift along with them.

Take, for example, the case of municipal officials who came to disconnect water within an informal settlement to the west of Durban. When they were seen by Sipho, a local resident in the settlement, he asked what they were doing. The workers apparently explained that they had been instructed to come because 'people are stealing water here'. Sipho, however, contested this claim by asking, 'Whose water are they stealing?' Sipho argued that the water they had tapped was rightfully theirs: it had been 'allocated' to them by the municipality. By this, he meant that the pipework fed the settlement's ablution block, which housed taps providing residents with free water. Sipho was claiming that the tapping of pipes from this original infrastructure was a legitimate extension of this state allocation. Whilst the workers agreed, they still had to report the disconnection to their line managers. From the municipality's perspective, household pipelines should be professionally installed and metered. The group reached a compromise: In order to respect the permissive space of the residents *and* the occupational orders of the workers, the pipe was disconnected, photographed and reconnected by the community 'in front of them' before the hole was refilled. 'They never came any more', Sipho concluded, 'I don't have a problem with water, up to today. Because they went back to their seniors [and] they explained the concern of the community. It is our water, it is allocated to us. Now you are saying that we are stealing it, but it is ours'.[16] In this instance, Sipho and his neighbours had successfully negotiated the informal fulfilment of a moral economy of citizenship, against the law and despite the state. Of course, not all negotiations were so successful. Across eThekwini – particularly during the decade from 2002 when the eThekwini municipality barred formal electricity from informal settlements – many settlements experienced police raids during which their wires were disconnected and their cables confiscated. Even in such cases, however, prosecution was rare. In other words, the permissive space in which residents were operating may have been disrupted or temporarily closed, but it did not fully collapse. Indeed, their claims to resources and permissive space were resurrected with the wires that residents re-acquired and reconnected across the settlement, often within hours of the police taking them down (Chance 2018). Claims on the moral economy of citizenship – entangled as they sometimes are with other claims – play an important role in people's everyday lives, even when they are illegal.

Conclusion

In this chapter, I have argued that the notion of the 'moral economy' can be utilised outside the context of the eighteenth-century crowd: the ways in which people negotiate access to social goods beyond the market is of interest across time and space. That said, some adjustments to Thompson's (1971) formulation must be made: First, economic shifts are rarely as clear-cut as Thompson's work suggests. That said, battles over moral economies are not limited to moments of extreme economic change. Second, moral economies need not be conservative: they may be pre-figurative. In this chapter, my focus has been on the moral economy of citizenship that emerged during the transition from apartheid, grounded in constitutional law. State-citizen relations were not being conserved through law; they were to be remade by it. This point is significant, I have argued, because pre-figurative moral economies give rise to a very particular form of politics – the 'politics of patience'.

Understanding constitutional rights as a moral economy of citizenship enables us to understand the important role that the politics of patience has played in post-apartheid South Africa. That, in turn, gives us certain insights into the relationship between statehood and citizenship since 1994. Ultimately, however, the concept is limiting because it encourages us to overlook the important ways in which statehood has always been a pluralised project and the law has always been normatively fickle. Realising this, we need to look beyond a formal moral economy of citizenship to other moral economies that people are trying to negotiate, which may be forged against the law, not through it.

Notes

1 Fumanti and Werbner (2010: 5) also allow for 'novel' ideas and practices constituting moral economies – the focus here on 'prefigurative' approaches is more deliberately future-oriented, grounded in a more systematic vision of a just future.
2 As I explain later, I use this term in a different sense to Appadurai (2001).
3 This develops a phrase in the South African context that has taken varied and diverse forms elsewhere, ranging from citizenship claims in the African diaspora (Fumanti and Werbner 2010) to discursive practices in China (Smart and Smart 2017) and the subjectivities of workers in post-war Poland (Lebow 2010). A moral economy that rests on citizenship is also, of course, highly exclusionary (Neocosmos 2008). I do not have the space here to explore the ways in which this promised provisioning has fed into a broader 'political economy of xenophobia' (Misago 2017) save to note its significance (Steinberg 2012).
4 This generalises Nthambeleni's (2008) argument, which focused on the politics of SANCO members. Nthambeleni argued that whilst there was an attempt to reconceive SANCO as an institutional support for the ANC in government, local branches were critically engaged and sometimes demonstrated active resistance. 'The development of a new grassroots resistance identity', he concludes, 'could be glimpsed in the 1990s but became more pronounced in the 2000s' (Nthambeleni 2008: 230).
5 I draw this personal observation from the meetings that I attended during eight weeks of research in 2008 and during observations at meetings in April 2019, as well as sporadic meetings in intervening years.
6 There is insufficient space in this article to explore the important questions of migration, being and belonging beyond citizenship.
7 Had I more space, for example, I would delve into the degree to which the importance of citizenship (membership to the state) has been blurred or displaced by the importance of partisanship (membership to a particular political party).

8 Legislation surrounding electricity in South Africa means that a right to access electricity exists even if there is no 'absolute right to electricity' itself (http://www.schindlers.co.za/2016/free-basic-electricity-do-you-have-a-right-to-it/).
9 As I argued earlier, however, the incorporation of market logics will never be absolute. This is why we can speak of differently moralised economies, not demoralised economies.
10 This is a pseudonym, used to protect the interviewee's identity. This interview was conducted as part of my PhD project (Cooper-Knock 2014) and was conducted in Durban in August 2010.
11 Brava Media. n.d., 'Electricity Is Life', last accessed 28 July 2020, https://vimeo.com/130205868.
12 In this context, authorisation refers to official authorisation by the state.
13 For more on the legal parameters of eviction, for example, see SERI (2015).
14 An isiZulu word for 'snakes' – the term 'izinyoka-nyoka' is used to refer to informal electricity connections, and the term 'izinyoka' can also refer to those who are involved in cable theft and informal electricity connections.
15 Legitimacy, in this context, is taken to be a multi-faceted, socially embedded claim that is under constant negotiation.
16 Interview with author, Durban, 21 November 2018. This was part of an ongoing ESRC project exploring the politics of fire within informal settlements.

References

Appadurai, A. 2001. 'Deep Democracy: Urban Governmentality and the Horizon of Politics', *Environment and Urbanization* 13(2): 23–43.

Arnold, T.C. 2001. 'Rethinking Moral Economy', *The American Political Science Review* 95(1): 85–95.

Auyero, J. 2012. *Patients of the State: The Politics of Waiting in Argentina*. London: Duke University Press.

Barchiesi, F. 2011. *Precarious Liberation: Workers, the State, and Contested Social Citizenship in Postapartheid South Africa*. Albany: State University of New York Press.

Biko, S. 1978. 'Towards True Humanity in South Africa', *The Ecumenical Review* 30(4): 355–368.

Bond, P. 2001. *Elite Transition: From Apartheid to Neoliberalism in South Africa*. London: Pluto Press.

Brava Media. n.d. 'Electricity Is Life'. Last accessed 28 July 2020. https://vimeo.com/130205868.

Brogden, M., and P. Nijhar. 1998. 'Corruption and the South African Police', *Crime, Law and Social Change* 30(1): 89–106.

Carswell, G., Chambers, T. and De Neve, G. 2019. 'Waiting for the State: Gender, Citizenship and Everyday Encounters with Bureaucracy in India', *Environment and Planning C: Politics and Space* 37(4): 597–616.

Chance, K.R. 2018. *Living Politics in South Africa's Urban Shacklands*. Chicago: University of Chicago Press.

Chaskalson, A. 2003. 'From Wickedness to Equality: The Moral Transformation of South African Law', *International Journal of Constitutional Law* 4(1): 590–609.

Christiansen, E.C. 2007. 'Adjudicating Non-Justiciable Rights: Socio-Economic Rights and the South African Constitutional Court', *Columbia Human Rights Law Review* 38(2): 321–386.

Cooper-Knock, S.J. 2014. 'Everyday Policing, Statehood and Sovereignty in South Africa: A Case Study of Responses to Theft and Robbery in eThekwini', DPhil Thesis. Oxford: University of Oxford.

Cooper-Knock, S.J. 2018. 'Beyond Agamben: Sovereignty, Policing and 'Permissive Space' in South Africa, and Beyond', *Theoretical Criminology* 22(1): 22–41.

Dawson, H. 2014. 'Youth Politics: Waiting and Envy in a South African Informal Settlement', *Journal of Southern African Studies* 40(4): 861–882.

Desai, A. 2002. *We Are the Poors: Community Struggles in Post-Apartheid South Africa*. New York: Monthly Review Press.

Dobler, G. 2020. 'Waiting: Elements of a Conceptual Framework', *Critical African Studies* 12(1): 10–21.

Dugard, J. 2009. 'Power to the People? A Rights-Based Analysis of South Africa's Electricity Services', in D. McDonald (ed.), *Electric Capitalism: Recolonising Africa on the Power Grid*. London: Routledge, pp. 264–287.

Fumanti, M., and P. Werbner. 2010. 'The Moral Economy of the African Diaspora: Citizenship, Networking and Permeable Ethnicity', *African Diaspora* 3(1): 2–11.

Gallens, M. 2017. '"Patience Is a Virtue", Mantashe Tells Soweto Residents Who Have Waited 22 Years for Houses'. Last accessed 20 June 2019. https://www.news24.com/SouthAfrica/News/patience-is-a-virtue-mantashe-tells-soweto-residents-who-have-waited-22-years-for-houses-20170106.

Gibson, N. 2016. 'The Ethical Struggle to Be Human: A Shack Dwellers Movement in South Africa'. *Heinrich Böll Stiftung: The Green Political Foundation* (blog). Last accessed 20 June 2019. https://www.boell.de/en/2016/01/19/ethical-struggle-be-human-shack-dwellers-movement-south-africa.

Götz, N. 2015. '"Moral Economy": Its Conceptual History and Analytical Prospects', *Journal of Global Ethics* 11(2): 147–162.

Greenberg, S. 2009. 'Market Liberalisation and Continental Expansion: The Repositioning of Eskom in Post-apartheid South Africa' in D. McDonald (ed.), *Electric Capitalism: Recolonising Africa on the Power Grid*. London: Routledge, pp. 73–108.

Gumede, W.M. 2008. *Thabo Mbeki and the Battle for the Soul of the ANC*. London: Zed Books.

Gupta, A., and A. Sharma. 2006. 'Introduction: Rethinking Theories of the State in an Age of Globalisation', in A. Gupta and A. Sharma (eds.), *Anthropology of the State: A Reader*. Oxford: Blackwell Publishers, pp. 1–42.

Hagmann, T., and D. Peclard. 2010. 'Negotiating Statehood: Dynamics of Power and Domination in Africa', *Development and Change* 41(4): 539–562.

Hall, S. 1996. *Race: The Floating Signifier.* Media Education Foundation. Last accessed 20 June 2019. https://www.mediaed.org/transcripts/Stuart-Hall-Race-the-Floating-Signifier-Transcript.pdf.

Hansen, T., and F. Stepputat. 2001. 'Introduction: States of Imagination', in T. Hansen and F. Stepputat (eds.), *States of Imagination*. New York: Dukes Press, pp. 1–40.

Hart, G. 2014. *Rethinking the South African Crisis: Nationalism, Populism, Hegemony*. Georgia: University of Georgia Pres.

Jeffrey, C. 2008. 'Waiting', *Environment and Planning D: Society and Space* 26: 954–958.

Joseph, G., and D. Nugent. 1994. *Everyday Forms of State Formation: Revolution and the Negotiation of Rule in Modern Mexico*. Durham, NC: Duke University Press.

Justice and Correctional Services. 2015. 'Criminal Matters Amendment Bill: Public Hearings'. Parliamentary Monitoring Group. Last accessed 20 June 2019. https://pmg.org.za/committee-meeting/21565/.

Khunou, G. 2002. '"Massive Cutoffs": Cost Recovery and Electricity Service in Diepkloof, Soweto', in D.A. MacDonald and J. Pape (eds.), *Cost Recovery and the Crisis of Service Delivery in South Africa*. Cape Town: Human Sciences Research Council Publishers, pp. 61–77.

Lebow, K. 2010. '"We Are Building a Common Home": The Moral Economy of Citizenship in Postwar Poland', in F. Biess and R.G. Moeller (eds.), *Histories of the Aftermath: The Legacies of the Second World War in Europe*. New York: Berghan Books, pp. 215–230.

Lombard, M. 2013. "Struggling, Suffering, Hoping, Waiting: Perceptions of Temporality in Two Informal Neighbourhoods in Mexico', *Environment and Planning D: Society and Space* 31(5): 813–829.

Makhulu, A.-M. 2015. *Making Freedom: Apartheid, Squatter Politics, and the Struggle for Home*. Durham, NC: Duke University Press.

Mandela, N. 2007. 'An Ideal for Which I Am Prepared to Die – Part Two'. *The Guardian* (blog). Last accessed 23 April 2019. https://www.theguardian.com/world/2007/apr/23/nelsonmandela2.

Mandela Park Backyarders. 2010. Backyarder Arrested But Workers on Housing Site Agree to Suspend Construction. Last accessed 20 June 2019. http://abahlali.org/taxonomy/term/mandela_park/mandela_park/page/7/.

Mandyoli, S.S. 2012. 'Selected Speeches of Robert Sobukwe and a Mini-Biography'. Last accessed 23 April 2019. https://ilizwe.files.wordpress.com/2012/09/speeches-of-r-m-sobukhwe.pdf.

Mbali, M. 2013. *South African AIDS Activism and Global Health Politics*. New York: Palgrave Macmillan.

Misago, J.P. 2017. 'Politics by Other Means? The Political Economy of Xenophobic Violence in Post-Apartheid South Africa', *The Black Scholar* 47(2): 40–53.

Mujere, J. 2019. 'Unemployment, Service Delivery and Practices of Waiting in South Africa's Informal Settlement', *Critical African Studies* 12(1): 65–78.

Naidoo, P. 2010. 'The Making of "the Poor" in Post-Apartheid South Africa: A Case Study of the City of Johannesburg and Orange Farm', PhD Thesis. Durban: University of KwaZulu Natal.

Naidoo, P., and A. Veriava. 2009. 'From Local to Global (and Back Again?): Anti-Commodification Struggles of the Soweto Electricity Crisis Committee', in D. Macdonald (ed.), *Electric Capitalism: Recolonising Africa on the Power Grid*. London: Human Sciences Research Council, pp. 321–337.

Nefale, M. 2004. *A Survey on Attitudes to Prepaid Electricity Meters in Soweto*. Johannesburg: CALS.

Neocosmos, M. 2008. *From Foreign Natives to Native Foreigners: Explaining Xenophobia in Post-apartheid South Africa*. CODESRIA.

Nthambeleni, N.B. 2008. 'The South African National Civic Organisation: A Two-Tiered Social Movement', PhD Thesis. Johannesburg: University of Johannesburg.

Oldfield, S., and S. Greyling. 2015. 'Waiting for the State: A Politics of Housing in South Africa', *Environment and Planning A: Economy and Space* 47(5): 1100–1112.

Pithouse, R. 2014. 'An Urban Commons? Notes from South Africa', *Community Development Journal* 49(1): 131–143.

RSA. 2015. *Criminal Matters Amendment Act 18 of 2015*. Last accessed 28 July 2020. https://www.gov.za/documents/criminal-matters-amendment-act-18-2015-15-dec-2015-0000.

RSA & Ors v Grootboom & Ors. 2000 (11) BCLR 1169. (CC).

Ruiters, G. 2018. 'Municipal Indigent Programmes in South Africa', *Journal of Asian and African Studies* 53(2): 169–186.

Scott, J. 1977. *The Moral Economy of the Peasant: Rebellion and Subsistence in Southeast Asia*. New Haven, CT: Yale University Press.

Smart, A., and J. Smart. 2017. 'Judging Publics and Contested Exclusion: The Moral Economy of Citizenship in China', in S. Brandtstädter and H. Steinmuller (eds.), *Popular Politics and the Quest for Justice in Contemporary China*. New York: Routledge, pp. 74–90.

Socioeconomic Rights Institute [SERI]. 2015. *Resisting Evictions in South Africa: A Legal and Practical Guide*. Last accessed 20 June 2019. http://www.serisa.org/images/SERI_Dear_Mandela_Evictions_Guide_FINAL_FOR_WEB2.pdf.

Steinberg, J. 2012. 'Security and Disappointment: Policing, Freedom and Xenophobia in South Africa', *The British Journal of Criminology* 52(2): 345–360.

Thompson, E.P. 1971. 'The Moral Economy of the English Crowd in the Eighteenth Century', *Past and Present* 50: 76–136.

Thompson, E.P. 1993. *Customs in Common: Studies in Traditional Popular Culture*. New York: The New Press.

Wacks, R. 1984. 'Judges and Injustice', *South African Law Journal* 101: 266–285.

Walker, C. 1991. *Women and Resistance in South Africa*. Cape Town: David Philip.

Wedeen, L. 2003. 'Seeing Like a Citizen, Acting Like a State: Exemplary Events in Unified Yemen', *Society for Comparative Study of Society and History* 45: 680–713.

4 Fractured social contracts

The moral economy of protest and poaching in Cape Town, South Africa

Fiona Anciano

Introduction

At the southernmost point of Africa, on the edge of urban Cape Town, sits the small fishing community of Hangberg. Walking through the steep roads of this mountainside suburb, you will see evidence of the role that the sea plays in providing livelihood opportunities for the residents; from the harbour and fishmeal processing plant at the entrance of the suburb to the piles of abalone shells lying against the walls of informal shacks. Hangberg is a community steeped in fishing. The sea has provided a means of subsistence and income for generations of residents. In the modern state, however, to fish, you need a permit and a quota allocation from the national government. This is a luxury many residents do not possess, and so they fish illegally – poaching in their own bay. They do not only 'quietly encroach' (Bayat 2010) on the margins of illegality; they also protest against the state and against what they feel is an unfair allocation of resources in relation to their rights.

The historic democratic elections of 1994 and the subsequent promulgation of the South African Constitution embodied both the promises of a state committed to redressing the socioeconomic and political inequities of apartheid, and a new social contract (Ndinga-Kanga et al. 2020). The post-colonial developmental state pledged to address racial inequality, provide fairer access to livelihoods, deepen participatory governance and accountability, and defend dignity. For many citizens, these promises have not been fulfilled. Using a moral economy lens, this chapter will address the evolving social contract in post-apartheid South Africa, showing how the perceived failure of the state to uphold aspects of this social contract leads to a variety of responses, including protest and poaching. Why have fishers,[1] and those who support their demands, taken to the streets? What informs their desire to demand better access to fishing as a livelihood, and what does this tell us about the nature, and fragmentation, of the social contract in post-apartheid South Africa?

In Thompson's work on moral economy, riots or popular protests were situated in a historical context where authorities held a paternalistic role and were seen as responsible for providing for needs in times of crisis or shortage (Siméant 2011). This framing provides a useful starting tool to analyse discontent. However, to fully understand the drivers behind contemporary protests, the framework must be expanded to accommodate contexts where authorities are expected to facilitate, rather than provide, access to resources or needs. This chapter will expand on, and unpack, the idea of authority in relation to demands for access to subsistence or social goods and livelihoods. Applying an expanded moral economy framework can assist us in understanding why and how communities engage with the state and other public authorities when demanding their right to a livelihood.

After framing moral economy, this chapter will tell the story of the 2017 fishing protest in Hangberg, Hout Bay. Using an expanded moral economy frame, it is possible to unpack which political and social values underpinned this protest, reflecting both the 'moral' legitimizing notions behind the protest as well as behind the widely accepted practice of poaching. Bound up in the moral economy of Hangberg is a tense and complicated social contract. The chapter will then look at how this is reflected in relationships with state and non-state public authorities, who are expected to provide access to fishing rights but are reviled for the way in which they do so. The chapter argues that both moral legitimizing notions and public authorities in moral economies are not homogenous but multifaceted, reflecting the multiple and contending forms of governance experienced by citizens in the post-apartheid state. This sheds light on the nature of the fracturing social contract between a fishing community and the authorities it depends on to uphold the developmental promises of freedom and rights entrenched in the 1996 South African Constitution.

Expanding the moral economy framework

Writing and research on moral economy is well suited to reveal why some communities experience a perceived breakdown in the social contract with the state and other authorities when they are not able to access subsistence goods, and why they challenge this through, for example, protest (and other forms of illegal behaviour such as poaching). For Bohstedt (2016: 9), the idea of moral economy has two key components. First is a deeply felt right, on the part of protestors, to an affordable provision or good such as food. Second is a reciprocal social contract where it is the responsibility of authorities to underwrite popular subsistence necessities. Hossain and Kalita's (2014: 4) definition of moral economy as including '(1) conceptions of material justice, (2) the idea of clear limits to the tolerance of exploitation and unfairness, and (3) the impacts of sweeping economic and political change on understandings of fairness and equity' is also a helpful starting point if one is to understand the logics behind why residents in Hangberg poach and protest. A rich scholarship on protest in South Africa talks to the multiple drivers behind citizens taking to the streets and shows where there are demands for justice, equity, access to livelihoods and fairness (Lancaster 2018).

In the context of contemporary research in post-colonial societies, it is helpful to widen the traditional focus of moral economy research from a deeply felt right to a good to include the right to access to livelihoods or government services. Similarly, the reciprocal social contract with authorities can include governing elites and the ruling class, but it could also be extended to authorities such as brokers, mediators and patrons. This aligns with Zahar and McCandless's (2020) research arguing that a resilient social contract requires high levels of inclusiveness, both at the political settlement stage and thereafter. While it is important to be aware of conceptual stretching (Sartori 1970), this broader understanding of moral economy is consistent with research in South America (Auyero 2007) and Africa (Lonsdale 1992) and does not detract from the essence of the concept.

Thompson's work on moral economy was based on a period of historical transition and economic upheaval, reflected in the Industrial Revolution and the advent of the market (Siméant 2011). In the South African context, there is indeed ongoing historical and economic upheaval; however, the state is often not a central player in providing for needs. There are multiple and contending forms of governance in any single context, including bureaucratic (state led), developmental (state led), network (often civil society led) and informal governance (Anciano and Piper 2019). This echoes Lund's discussion of public authority as involving twilight institutions that are not in the state but mirror certain roles that the state theoretically plays. He argues that 'while government institutions are important, the state qualities of governance – that is, being able to define and enforce collectively binding decisions on members of society – are not exclusively nested in these institutions. A wider variety of institutions are at play' (Lund 2006: 685).

A moral economy frame that broadens the idea of public authority to include state and non-state actors is thus useful when analysing why residents protest and towards whom they direct their perceptions of a broken social contract. Often political parties, such as the ruling African National Congress (ANC) in South Africa, may play a significant role as a public authority. Looking at the relationship between citizens (who demand goods, access to livelihoods and services) and public authorities (such as political party representatives) is thus closely linked to theories such as clientelism and neo-patrimonialism. Both are based on the logic that authorities are responsible for a social contract and can and should provide, or at least manage access to, resources. Auyero (2007), writing on Argentinian protests, links moral economy and agency to clientelism. He acknowledges the 'grey zone' of the relationships between agents of collective action and authorities. Auyero's framing also points to the importance of not automatically drawing a distinction between demand-side agents (protestors) and supply-side agents (authorities), as these actors can play multiple roles, particularly in terms of brokers or mediators (see Piper and von Lieres 2014). The research in this chapter employs a broad understanding of public authority, linked to the notion of multiple and contending forms of governance beyond state institutions, to unpack the relationship between fishers and the perceptions of a fracturing social contract in post-apartheid South Africa.

Protest in the bay

It is a quiet Tuesday morning and I am sitting at a small coffee shop on a main street in Hout Bay, Cape Town, interviewing a local resident about her views on river pollution and urban governance. As we sip our coffee, a busload of tourists pour into the little shop. This is not unexpected in this beautiful bay at the foot of the world-renowned Chapman's Peak mountain. However, these tourists look shaken. And then we hear it – the sound of sirens and the low boom of small explosions. A few hundred meters away, where smoke billows, the residents of Hangberg, predominantly from the fishing community, are protesting. A tourist

Figure 4.1 Map of key settlements of Hout Bay.

bus was caught in their way as they lit fires on a main road running next to the working harbour. Protestors are unhappy about the allocation of small-scale fishing permits and have decided to take their grievances public, in an attempt to influence national policy makers.

The protests did not erupt without warning, and were not limited to fishing concerns. There were many indications that Hangberg residents were frustrated and feeling powerless. On 18 August 2017, a smaller housing protest started when the Hangberg community learnt that R100 million ($8 million) was earmarked for housing development and upgrading of the nearby Imizamo Yethu informal settlement (see Figure 4.1). They were upset as they felt they had waited years for title deeds and housing upgrades, and thus: 'We the women of Hangberg decided … that we would strike today' (Payne 2017). The Mayor did not visit the protestors as demanded, and they felt little attention was given to their plight (Fisher A 2018).

Several weeks later, the National Department of *Agriculture, Forestry and Fisheries* (DAFF) issued a proposal that the total allowable catch of West Coast rock lobster be reduced by around 69% from 110 kg to 50 kg for the six-month fishing season. This proposed reduction would have a significant effect on the livelihoods of many Hangberg residents. As one fisher explained, 'How can we live with 50 kg? That's probably R9000 [$630] for six months, and the next six months? How are we going to survive?' (Jordan and Hyman 2017). The fishers in Hangberg held a meeting to discuss the proposed cuts and the 'majority wanted to protest' (Fisher A 2018). A local fisher explains how he tried to persuade them to wait a few days for DAFF to respond to a memorandum of concerns that was sent to the Minister of DAFF, but to no avail. Two days later, on 12 September 2017, 'there was an uprising' (Fisher A 2018).

The protest itself started relatively peacefully, with the road to the harbour being blocked and tourist buses being encouraged to leave. However, when the protestors felt they were not getting a response from DAFF or other government representatives, they escalated their methods, using petrol bombs and setting alight bins. 'We were waiting for de Lille [the Mayor of Cape Town] or DAFF and didn't get a response for hours' (Fisher A 2018). A separate breakaway group also attempted to rob two tourists, and the windows of a nearby restaurant were stoned. Furniture was stolen and placed on the fires (Jordan and Hyman 2017). As a fisher at the scene explained, 'the police then had to intervene. They said it had turned from a protest to a riot because of the private property' (Fisher A 2018). Police used stun grenades, a water cannon and rubber bullets in what became a running battle along the harbour and up to the entrance of Hangberg. After the first day of protest, the Deputy Minister of Police agreed to an *imbizo* (meeting) with the community, but the meeting was cancelled. When the protestors heard this, they set a construction vehicle alight. A police van was also petrol bombed. Several protestors said that they would continue burning things until they were addressed by a government minister (Furlong et al. 2017). The protests and running battles with the police continued on and off for three days (see Figure 4.2).

Figure 4.2 Protestors speaking during a fishing and housing protest, September 2017 (Sullivan 2017).

Values formation: Legitimizing notions in the moral economies of protest and poaching

Protestors in Thompson's example of food riots were triggered by grievances that operated within a popular consensus as to what practices (baking, milling, etc.) were legitimate and illegitimate. These grievances and the allied consensus to protest were based on a 'consistent traditional' view of social norms. The riots reflected a 'vision of a better world', and the crowd's activities were shaped by a legitimizing notion: the 'men and women in the crowd were informed by a belief that they were defending traditional rights and customs' (Thompson 1971: 78–9). In a similar way, the protests in Hangberg were triggered by a set of legitimizing notions, including the right to fish, the right to agency and the right to access a market to sell their goods. When legitimizing notions, or sets of values, are undermined by state policy and law enforcement, there is a resultant perception that the state, and its agents, have not upheld a perceived social contract.

Custom, rights and fish as a social good

Fishing (and associated practices) as a form of custom and livelihood is woven into the historical fabric of Hangberg. This 'traditional right' is based on customs passed down through generations. Traditional fishing practices in fact predate the formal establishment of Hangberg as a residential area. Between the 1950s and 1980s, industrial fishing and processing activities formed the foundation of the local economy in Hangberg. A large proportion of the population worked in the fish factories near the harbour and on industrial fishing vessels (Schultz 2015). An established fisher describes how 'my mother and my grandmother worked Hangberg into the fishing village it is today' (Fisher B 2018). Indeed, history shows that fishing, particularly for rock lobster and abalone,[2] is woven into the social fabric of the Hangberg community (Hauck 2009). In a focus group conducted with poachers in Hangberg, it was clear that they all come from families steeped in the history of fishing off the Hout Bay coast (Poachers 2016). Another fisher explained, 'I'm on the sea for more than twenty years. I grew up on the seas. I'm a fisherman all the years … my father's been a fisherman … my mother worked in the processing plants' (Fisher C 2016).

Interlaced with traditional fishing practices is a strong belief that fishers have the right to fish in the sea – with or without government permits. This legitimizing notion gave moral authority to the protestors, reinforcing their agency. They were motivated by an inherent sense of justice. As one fisher explained, the protestors did not spend a lot of time trying to negotiate with authorities because 'We cannot negotiate for something that is ours'. He continued, 'we have a human right to the sea … the right to sustain your family from the sea' (Fisher A 2018). Echoing the idea of a social contract formed at the start of democracy in 1994, a second fisher clarified that the right is legitimized by the notion that 'this is our food … on our table'. It is also a legal right: 'Citizens have the right to resources according to the Constitution' (Fisher B 2018). Indeed, the right to customary fishing practices, in certain contexts, has recently been upheld by the Supreme Court of Appeal in a case brought by the Dwesa-Cwebe community. In a landmark judgement, the court ruled that there was 'extensive evidence' of a customary law system regulating all aspects of life, including access to and use of marine resources in the community. If knowledge of this system was passed from generation to generation and fishing practices were done sustainably, then the court ruled the community can fish in a marine protected area without a permit (Ampofo-Anti 2018).

Thomas Arnold (2001) argues that reconceptualising the moral economy in terms of social goods can reveal further grounds for politically significant moral indignation. He conceptualizes 'social goods' quite broadly. Drawing on a justice-based application, he sees even goods normally considered pure commodities as social goods, for they 'consist of shared understandings about the beneficial characteristics attributed to a given object' (Arnold 2001: 91). People are often placed in specific social, economic and even political relations based on the patterned use, display and exchange of social goods. While Arnold (2001) refers to social goods such as cattle or rice as key to a moral economy analysis, this framing also sheds light on the fishing protests, where fish are understood as an individual and communal social good. Intrinsically, communal social goods are sources for shared notions of legitimacy, and thus popular consensus about what is a legitimate or illegitimate action is vested in the nested meaning of specific social goods. A threat to this good can prompt an organized response that may range from public criticism to violent forms of protest (Arnold 2001). Three days of active (and occasionally violent) protests were broadly seen as legitimate by most in the Hangberg community because of the inherently cultural and economic properties of fish, and the role it plays, as a social good, in structuring identity and relationships in Hangberg. The idea of fish as a social good is also a legitimizing notion behind the right to fish illegally (popularly known as poaching or protest fishing, depending on which side of the socio-economic and ideological spectrum one sits). When access to social goods, including the right to eat and sell fish, are denied by state policy or law enforcement, this can trigger a material breakdown in the social contract between citizens and the state.

Racial exclusion and powerlessness as a legitimizing frame

Coupled with the understanding that the protests were legitimized by fish as a social good, and by a perceived right to the sea, is a complex race narrative entwined with South African political history. The vast majority of Hangberg residents are classified as part of the coloured race group. This is a residual racial category invented for people who, according to the 1950 Population Act, were 'not a white person or a native' (Union of South Africa 1950). Genetic studies suggest that some 43% of coloured people are descended primarily from the indigenous KhoiSan (de Wit et al. 2010). Indeed, Hangberg was designated a coloured area to provide labour for the fishing industry (SA History Online 2015). From 1950 to

1980, there were a series of forced removals where coloured people were moved from the Hout Bay village and the Hout Bay valley to Hangberg (Sowman and Gawith 1994). Many of the fishers spoken to (Fisher A 2018; Fisher B 2018) hold the perception that 'we were not white enough under apartheid' and now, under a black-led ANC government that pursues race-based affirmative action policies, we 'are not black enough' (see also Anciano 2014: 18). This is voiced clearly by poachers who explain that, 'we were the ones who fought apartheid with our brothers ... now it's like we don't exist or something ... they are marginalising the coloured people' (Poachers 2016). Racial equality, however, was a central tenet of the Constitution and the concomitant social contract underpinning the political settlement in 1994.

In an incisive article outlining the evolution of abalone poaching in South Africa, Steinberg (2005: 2, 6) notes that the most 'interesting and important' reason for the rise of an illicit market in abalone trade is the 'mutation in the socio-political identities of the coloured fishing communities'. The evolution of a distinctive political consciousness has motivated both the illegal fishing of the abalone stock, and arguably protests against not being given permits to fish. Talking about the years after the fall of apartheid, Steinberg (2005: 6) argues that 'Given the politics of the moment, a great many people who had lived their lives on the coastline believed that they were entitled to it, and to a share of the benefits that accrued from harvesting it'. This belief is expressed as an entitlement to take fish (poaching) and a demand for permits (protest) because the coloured working class has been betrayed by both the apartheid state and now the ANC state, thus they are due historically denied social justice. Indeed, one poacher explains, 'in the apartheid years the government enriched you because you were white, now it's time for us. Give us what is our due' (Poachers 2016).

The formation of values in the moral economies of Hangberg can be understood on historical, cultural and racial grounds (as described earlier), or framed as a sense of powerlessness and a desire to establish agency in a context where fishers (and the broader community) feel helpless against the bureaucracy of the state. In an interview with a fisher and community leader (Fisher B 2018), he describes how 'most people who live here like to think of themselves as victims, forgotten by the state, it is a community mindset'. He describes how this perspective, coupled with 'years of frustration', led to the three-day protest. The 2017 protest is indeed one in a number of significant protest moments in Hangberg. In 2010, residents fought against the City of Cape Town's attempt to evict residents living in informal housing above a firebreak on the mountain. This became known as the 'Battle of Hangberg' and resulted in a violent running battle with the police over several days (Anciano and Piper 2019: 102).

The community leader (Fisher B 2018) further notes that 'the youth in particular feel they have no power, this was a moment when they could fight the police, fight the state, have some authority'. The feeling of powerlessness and a desire to express agency are indeed also echoed in the 'not white enough, not black enough' sentiment and can be seen as underpinning values that alter and affect the overall moral economy of Hangberg. The frustrations of the youth in Hangberg echoes other research on protest in South Africa that describes how many of the youth who participate in violent protest are unemployed, live in poverty, and see no prospect of a change in their circumstances:

> Theirs, they feel, is a half-life, as they are unable to participate as full citizens in the economy and society. For young men in particular protest gives them an opportunity to exert their masculinity and to be seen as representing the community.
>
> (von Holdt et al. 2011: 3)

This sense of powerlessness points again to a fracturing social contract. Unemployed youth have no access to livelihoods and cannot fully engage as citizens with rights and responsibilities. Instead, the feelings of subjecthood lead to retaliation against the state, reflecting unfulfilled promises of the 1994 political settlement.

In concluding this section, it is important to note that when looking at values formation in a community such as Hangberg, it must not be presumed that these values are uniform or that there is one moral economy. As other chapters in this book argue (see Wheeler, Chapter 1), potential multiple moral economies occur within one context. In this case, while the majority of the Hangberg 'community' identified and supported a protest for expanded fishing rights, several felt this was not the right approach. Others, as described at the beginning of the chapter, were focused on the moral economy of housing and were less motivated by demands related to fishing quotas. Some, particularly youth, were motivated by a moral economy driven by the need to protest and rebel against the state, in a context of powerlessness, regardless of the social good or livelihood in question. Others were motivated by a sense of racial exclusion. In terms of poaching, there are elements of its operations many feel are not good for the broader community, such as the impact it has on young boys who work on the mountains at night carrying abalone shells and are then unable or unwilling to attend school (Tefre 2010). They also see the negative effects of younger men earning 'easy money', indeed more than their parents, and how this distorts the household balance of power (Fishers D and E 2016). Nonetheless, while many Hangberg residents understand that poaching is legally a crime, and it is certainly seen as a dangerous way of earning an income, it is also clearly one part of the moral economy in the community. Residents can also wear multiple hats, supporting different moral economies and values at the same time. It is important to note that a sense of what is just – or the 'moral' legitimizing notion – is not always the same, at the same time, for everyone in the settlement.

Public authority and the moral economy

When residents of Hangberg barricaded the entrance to Hout Bay Harbour and set alight furniture, they did this to attract attention – not from the tourists caught on the wrong side of the road, or from fellow Hout Bay residents, but from the state and other public authorities. Inherent in the moral economy of Hangberg is the belief that the state, and its allied representatives, must provide fair access to resources and opportunities for livelihoods. This resonates with the idea of a reciprocal social contract where it is the responsibility of authorities to underwrite popular subsistence necessities (Bohstedt 2016). Drawing on Lund's (2006) framing of public authority, this chapter uses public authority in a broad sense to refer to both the 'state' from above and 'twilight institutions' that include local and external institutions involved in the exercise of public authority. These may operate between state and society and between public and private. Here, 'public authority is the amalgamated result of the exercise of power by a variety of local institutions and the imposition of external institutions, conjugated with the idea of a state' (Lund 2006: 686). Public authorities are those whom Hangberg residents hold accountable and responsible for the provision and management of social goods, or access to livelihoods, in line with Thompson's idea of authority. It is when this relationship between public authorities (including the state) is perceived to be corrupt or poorly regulated that discontent erupts.

The state

While not always the central governance actor, the state is, however, an authority that can obstruct the attainment of resources or livelihoods. Inversely, the state is a central actor in terms of granting access to government services, goods or livelihood opportunities (such as trading licences or fishing permits). In this sense, communities have a strong normative understanding of how the state should uphold a social contract, even when it may at times be beyond its sphere of activity to do so. The democratic transition in South Africa, and the allied social contract, included a promise of fairer access to livelihoods and improved economic equality.

The state, in its incarnation as the National Department of Agriculture, Fisheries and Forestry (DAFF) is certainly a key actor that fishers target through protest. DAFF is a significant public authority and party to a social contract of accountability because it is responsible for policy regarding small-scale fishing (SSF), quota allocations and the management of permits. After court action in 2009, the Department was forced to create a new SSF policy, which became law in June 2012. A central objective of the new SSF policy is to recognize the socio-cultural, ecological and economic linkages of small-scale fishery systems. This approach, *in theory*, upholds constitutional promises of a developmental state, supporting the social contract between fishers and authorities. It promotes the sustainable use and management of marine resources while having a more human rights focus and prioritizing socio-economic development, food security and poverty alleviation (Masifundise Development Trust et al. 2012). An important aspect of the SSF policy is the stipulation that fishers must now form community-based legal entities (CBLEs) to claim rights for their communities (Stern 2013). The prerogative to issue these rights lies with the Minister of DAFF, and in this regard the Minister, who is part of the ruling ANC party, the government and the state, holds authority in the moral economy of fishing in Hangberg.

The perception that the state has fractured the social contract in small-scale fisheries governance is due in large part to disagreement around the allocation of fishing rights. As demonstrated by the Hangberg protests, the majority of small-scale fishers feel that fishing rights have not been granted in an equitable manner. The total allowable catch (TAC) favours industrial companies, leaving individuals with a disproportionally small proportion of the total allocation (Schultz 2015). This inequity is felt strongly at the local level, with the question of who is allocated rights in a community such as Hangberg causing tension and resultant distrust of national government.

Community-based public authority

While the national department has the final say in the allocation of fishing permits and quotas, the role of CBLEs is important. CBLEs are meant to be part of co-management structures, a formalized structure that serves as the main local management responsible for overseeing compliance, protecting community interests, and ensuring the sustainability and conservation of ecosystems and marine resources. A fisher can theoretically become a member only if they are an unemployed South African and can demonstrate daily or historical involvement in the fishing value chain (harvesting, processing and marketing) over a period of ten years. They need to show historical dependence on fishing, proof of subsistence on their catch and/or involvement in semi-commercial activity like barter or sale. These individuals will be identified by the community and are placed on a list verified by the Minister (or the Department or a third party), and then become a CBLE (Masifundise Development Trust et al. 2012).

The role of local leaders in community-based organizations such as CBLEs is significant when it comes to the verification of permit allocations. Fishers have to be seen by the state as 'bona fide' in their applications. Some local fisher representatives may have exploited this verification process. The local fisher leaders facilitate access to permits on condition that they are granted authority to market the permit holder's catch for a percentage of the sale price. This process leads to indiscriminate endorsement of as many permit holders as possible and complex relations in fishing communities with a local form of public authority – fisher representatives. DAFF has been unable to manage this verification process effectively, leading to the inequitable pattern of fishing rights allocation processes being repeated (Schultz 2015).

A related concern is that the communities affected by fishing policies have no means to monitor the application of the permit allocation, or to stop outsiders from illegally fishing in the waters near their communities. The SSF policy tries to deal with this by allowing for the training and appointment of a community monitor (albeit with unclear powers of arrest or punishment) and with the allocation of fishing rights to communities, not individuals (Stern 2013). Since communities need representatives, this creates a new layer of local-level public authority.

Although the SSF policy is *theoretically* a step in the right direction in supporting small-scale fishers, it is clear that many fishers in Hangberg are unhappy with how the policy is implemented and how they are treated by the state, and by their own local representatives. Many in the community are happy for residents who are seen as fishers (based on family histories and livelihood practices) to receive a quota, but where a community member gets a quota and is not historically a fisher, it is viewed as unacceptable, thus breaking the moral economy (Fisher A 2018; Fisher C 2016). The perceived inequity in the allocation of permits and quotas by DAFF (at times as a result of abuse of the system by local fisher representatives), is fracturing relationships in Hangberg between those who are beneficiaries and those who are not. The management of CBLEs and permits *in practice* is damaging a perceived social contract between the state and small-scale fishers.

Political actors as public authorities

There is a complicated relationship between political parties, their representatives and Hangberg fishers. The role of political parties and political affiliations was raised in almost every interview and focus group conducted in Hangberg, even where respondents were not asked about this. After 1994 there was a belief by several groups of fishers that collaborating with the ANC would lead to a better chance of getting a quota allocation (Fishers D and E 2016). Originally respondents felt the ANC was their main hope of support when the ANC wielded widespread power nationally and provincially (Poachers 2016; Fisher F 2015). This reflects an idea of the ANC as a key public authority, one which has responsibility for upholding the state's social contract. Indeed, voting patterns demonstrate that the ANC received the highest number of votes in Hangberg in the elections after 1994, before this pattern changed in the mid-2000s (Anciano and Piper 2019). As a member of a fishers' focus group explains, 'the people' of Hangberg tried to access opportunities through political parties 'back in the day … but it has stopped'.

Initially poachers believed that being affiliated to a political party that was in power would increase their chances of receiving fishing permits or influencing policy making. In the late 1990s, they started to work with the ANC through one of its local leaders. As they explain:

> All the things we did for the ANC … we done everything … so we decided since our story is taking so long that we are gonna hook up with [a local ANC leader], intermingle with him … through this intermingling we had to do a lot of things for the ANC … canvassing and getting votes for the ANC … I mean we did a lot of things for the ANC … no result! No result!'
>
> (Poachers 2016)

A more recent attempt to use politics to win influence can be seen through the harbour project in Hout Bay. Hout Bay has an active working harbour that is used by commercial fishers, small-scale fishers, tourist boats and poachers. Over time, a gentrification process has taken place in the harbour, with the main activities moving from fishing to tourism, prompting a rethink of the harbour design and use. Harbours are national mandates, controlled by DAFF and the Department of Public Works. Due to its substantial economic potential, the ruling party in the Western Cape province, the Democratic Alliance (DA), is very keen for provincial government to take control of the harbour (Zille 2017), which would move political influence away from the ANC.

A response to these contending political public authorities is seen in a family where the brother is a representative of the ANC-aligned Hout Bay Civic Association (HBCA), while the sister sits on a DA-aligned mediation forum. In this context, the brother feels that working with the ANC may be more beneficial in the current political climate as it could improve his ability to benefit from the harbour redesign. A colleague of his who works with him on an Aqua Farming project was very surprised to hear that he had joined the HBCA. He explained that this was 'news to me' but that it makes perfect sense politically:

> You must remember national government runs our natural resources, our seas, our fishing quotas and the minister comes from the [ANC-aligned] SACP [South African Communist Party] … And his deputy … is from the ANC and they have a very specific mandate and that is transformation in the fishing industry. Now these guys [the brother] didn't get a quota, he's a poacher … in other words … he decided to jump ship to the ANC with the hope that the ANC will give him a quota that's coming up … With the hope that with the ANC will come money and help to build this [aqua] farm'.
>
> (Fisher G 2015)

The flexible support for political parties that may assist in improving personal livelihoods is most starkly evident in a statement from an aging Rasta poacher (Poachers 2016). He describes with emotion how, even after he has suffered under apartheid, he would vote for any party that would support his ability to become a legitimate fisher – even if that party was conservative and white-led:

> If any party can come and say this is what we can give you, we just need to secure your vote … and I can see what they are offering is real and true … I'll vote for them yes … a hundred times … the EFF … even for the Freedom Front Plus.[3]

It is clear that the fishers we spoke to saw a dual benefit to supporting a political party. First, if the party won power (either locally, provincially or nationally), there was a hope that political representatives would use formal democratic channels to support the plight of illegal fishers by changing fishing policy, which would lead to the allocation of larger quotas for small-scale fishers.

The second benefit of supporting a political party is more closely related to the idea of clientelism than the idea of representative democracy. Here, respondents describe how they were promised direct benefits if they 'worked for' the ANC. When discussing how quotas are allocated and who ends up running co-operatives, one fisher explained that it is all about who you know that will determine if you get in:

> At the beginning they were 10 clever guys, these directors, who were not fishermen. They had all these connections with ANC officials … and they applied for the quota. However, they were turned down and told that they

> need 20 fishermen to succeed with their application. Then they came to approach us 20 fishermen.
>
> (Fisher H 2016)

Although the co-operative was successfully set up, over time the non-politically connected fishers were eased out of their role in the co-operative. Most signed their shares away, or remain in the co-operative on paper only, receiving few, if any, dividends (Fisher C 2016).

Just prior to the protests in 2017, the sense that the community was abandoned by political parties ran high. The social contract – which in South Africa implied greater democratic accountability and fairer access to livelihoods – was fractured. As one respondent noted:

> The DA rules here and they don't listen to the ANC. The ANC won't help us because we didn't vote for them ... The whole of Hangberg has no say in what happens in Hangberg. We have no impact on decisions. Resources are ruled by the ANC nationally and the DA rule the Western Cape. They make decisions and it filters down to us and we bear the brunt.
>
> (Fisher A 2018)

There are multiple public authorities in any one context, as demonstrated in the case of Hangberg. Here the state, local-level fisher representatives and political parties all hold authority to maintain the social contract for developmental fisheries governance. Yet, fishers, protestors and poachers broadly feel all forms of authority have let them down. As research has shown, where public authorities do not respond to peaceful engagement, violent protest becomes more acceptable as this is seen to yield more successful results (Bohler-Muller et al. 2017). In this sense, the moral economy of protest and poaching in Hangberg demonstrates the clear fracturing of a developmental social contract. It is also notable that government authorities are not the only, or even main, form of public authority in the moral economies of Hangberg. Political parties, particularly the ANC given its role as a liberation party, are seen as important actors.

Conclusion

The story of fishing protests and poaching in Hangberg reflects a community embedded in an overlapping set of moral economies. Understanding the drivers behind these moral economies illuminates why residents take to the streets and engage in illegal poaching – it is due to the legitimizing notions of social justice and rights to a sustainable livelihood bound up with reciprocal social contracts between a range of public authorities. Moral economy as a concept is thus helpful in pointing to the underlying causes of protest and political action (including quiet encroachment and more conventional forms of participation). It highlights both the sense of social justice embedded in historical and cultural norms and the embedded economy of a place or community. It also builds understanding of the different types of relationships actors (from an individual resident to a collective community) have with public authorities (from a national government department to a local broker), and how these can trigger or mediate protest and participation.

Moral economy as a theory, however, needs to be developed to allow for the unintended consequences of the role of the external (unembedded/nonculture) market and state policy. The role of the 'free' market complicates the moral economy framing where it distorts the economic gain from the use of a social good. In this case study, the profits to be made from abalone poaching have altered historical relationships with fishing. While fishing without a permit (technically poaching) has always been viewed as a legitimate part of the moral economy in Hangberg, the proceeds from sales to Chinese triads has brought new dimensions of criminality and inequality into the community (which, of course, may in turn lead to a new set of legitimizing notions). These high profits are not communally shared and so jar with Scott's (1976) notion of a 'subsistence ethic'.

It is also important to expand the use of moral economy to allow for ruptured binaries. There is not always one overriding legitimizing notion in any settlement. As this chapter demonstrates, there are a set of different, although often overlapping, values that underpin poaching and protesting in Hangberg. One person may hold multiple legitimizing notions at any one time, or different members of the settlement may associate with different values at different times. It is also necessary to fracture the binary of one public authority versus one homogenous group. The state is not the only authority in any given governance context. Again, this chapter highlights how a range of public actors can wield authority – from state departments to political parties to local fisher representatives. The state is often the most obvious and important public actor, but in South Africa, the liberation nationalism of the ANC (see Piper and Anciano 2015) imbues it with symbolic power as a public authority. In terms of fisher representatives, these brokers can at different times be associated with the state or be associated with the marginalized. Public authority is not homogenous or binary.

Finally, applying an expanded version of a moral economy framework highlights the nature and condition of the social contract in post-apartheid South Africa. The rise of protest and poaching is the manifestation of a fracturing, if not broken, social contract between the developmental and democratic promises of the 1996 Constitution and citizens unable to sustain a livelihood. Echoing the idea of a social contract formed at the start of formal democracy in the 1990s, fishers believe that the right to fish is legitimized by constitutional promises, which in their understanding stipulates the legal right to access resources to protect a livelihood (Fisher B 2018). The idea of a social contract is also premised on the view that the community of Hangberg is due historically denied social justice. Whites were enriched during apartheid and now, under a post-apartheid government, it is time for other racial groups to benefit from available resources. It is clear from the protests and illegal fishing in Hangberg that citizens feel these elements of the social contract have not been upheld.

Notes

1 Fishers in this chapter refers to both those who fish with legal permits and those who poach.
2 Abalone, also known as perlemoen (*Haliotis midae*), is a pale coloured form of large sea snail housed in a flat, shiny shell. It is traded live, frozen, tinned, canned or dried.
3 The Freedom Front Plus is a small conservative political party often associated with the South African white minority.

References

Ampofo-Anti, O. 2018. 'Court Gives Community Right to Fish without Permit'. *GroundUp*. Last accessed 25 June 2020. https://www.groundup.org.za/article/court-gives-community-right-fish-without-permit/.

Anciano, F. 2014. 'Non-Racialism and the African National Congress: Views from the Branch', *Journal of Contemporary African Studies* 32(1): 35–55.

Anciano, F., and L. Piper. 2019. *Democracy Disconnected: Participation and Governance in a City of the South*. London: Routledge.

Arnold, T. 2001. 'Rethinking Moral Economy', *American Political Science Review* 95(1): 85–95.

Auyero, J. 2007. *Routine Politics and Violence in Argentina*. Cambridge: Cambridge University Press.

Bayat, A. 2010. *Life as Politics: How Ordinary People Change the Middle East*. Stanford, CA: Stanford University Press.

Bohler-Muller, N., et al. 2017. 'Minding the Protest: Attitudes towards Different Forms of Protest Action in Contemporary South Africa', *South African Crime Quarterly* 62: 81–92.

Bohstedt, J. 2016. 'Food Riots and the Politics of Provisions from Early Modern Europe and China to the Food Crisis of 2008', *Journal of Peasant Studies* 43(5): 1035–1067.

Furlong, A., et al. 2017. 'Mayhem in Hangberg after Deputy Minister of Police Cancels Imbizo'. *SA People News*. Last accessed 25 June 2018. https://www.sapeople.com/2017/09/14/mayhem-hangberg-deputy-minister-police-cancels-imbizo/.

Hauck, M. 2009. 'Rethinking Small-Scale Fisheries Compliance: From Criminal Justice to Social Justice', PhD Dissertation. Cape Town: University of Cape Town.

Hossain, N., and D. Kalita. 2014. 'Moral Economy in a Global Era: The Politics of Provisions during Contemporary Food Price Spikes', *Journal of Peasant Studies* 41(5): 815–831.

Jordan, B., and A. Hyman. 2017. 'Hout Bay Protesters Go after Tourists, Boats, Restaurants and Helicopter'. *TimesLive*. Last accessed 15 June 2018. https://www.timeslive.co.za/news/south-africa/2017-09-12-hout-bay-protesters-go-after-tourists-boats-restaurants-and-helicopter/.

Lancaster, L. 2018. 'Unpacking Discontent: Where and Why Protest Happens in South Africa', *South African Crime Quarterly* 64: 29–43.

Lonsdale, J. 1992. 'The Moral Economy of Mau-Mau', in B. Berman and J. Lonsdale (eds.), *Unhappy Valley: Clan, Class and State in Colonial Kenya*. London: James Currey, pp. 265–504.

Lund, C. 2006. 'Twilight Institutions: Public Authority and Local Politics in Africa', *Development and Change* 37(4): 685–705.

Masifundise Development Trust, et al. 2012. *Small-Scale Fisheries (SSF) Policy: A Handbook for Fishing Communities in South Africa*. Retrieved 8 May 2018. http://toobigtoignore.net/wp-content/uploads/2014/05/SSFpolicyENG1.pdf.

Ndinga-Kanga, M., H. van der Merwe and D. Hartford. 2020. 'Forging a Resilient Social Contract in South Africa: States and Societies Sustaining Peace in the Post-Apartheid Era', *Journal of Intervention and Statebuilding* 14(1): 22–41.

Payne, S. 2017. 'Hangberg Residents Protest for Better Living Conditions'. *GroundUp*. Last accessed 15 June 2018. https://www.groundup.org.za/article/hangberg-residents-protest-better-living-conditions/.

Piper, L., and F. Anciano. 2015. 'Party over Outsiders, Centre over Branch: How ANC Dominance Works at the Community Level in South Africa', *Transformation: Critical Perspectives on Southern Africa* 87(1): 72–94.

Piper, L., and B. von Lieres. 2014. 'Introduction: The Crucial Role of Mediators in Relations between States and Citizens', in B. von Lieres and L. Piper (eds.), *Mediated Citizenship*. London: Palgrave Macmillan, pp. 1–22.

SA History Online. 2015. *Hout Bay*. Last accessed 26 January 2015. http://www.sahistory.org.za/places/hout-bay.

Sartori, G. 1970. 'Concept Misinformation in Comparative Politics', *American Political Science Review* 14(4): 1033–1053.

Schultz, O. 2015. 'Power and Democracy: The Politics of Representation and Participation in Small-Scale Fisheries Governance on the Cape Peninsula', PhD Dissertation. Cape Town: University of Cape Town.

Scott, J. 1976. *The Moral Economy of the Peasant Rebellion and Subsistence in Southeast Asia*. London: Yale University Press.

Siméant, J. 2011. '*The Concept of Moral Economy Applied to Riots and Protest in Poor Countries: How It Helps and Why It Should be Used with Caution – The Example of Mali*', *European Consortium for Political Research Conference*, Reykjavik, 25–27 August 2011. Reykjavik: University of Iceland.

Sowman, M., and M. Gawith. 1994. 'Participation of Disadvantaged Communities in Project Planning and Decision-Making: A Case Study of Hout Bay', *Development Southern Africa* 11(4): 557–571.

Steinberg, J. 2005. 'The Illicit Abalone Trade in South Africa', *Institute for Security Studies Paper 105*. Retrieved 8 May 2018. http://www.files.ethz.ch/isn/99200/105.pdf.

Stern, M. 2013. 'Sustainable Livelihoods and Marine Resources: How Does South Africa's Policy for the Small-Scale Fisheries Sector Consider Current Challenges on the Ground?', *SAIIA Occasional Paper No. 166*. Retrieved 8 May 2018. http://www.files.ethz.ch/isn/176493/saia_sop_166_%20stern_20131231.pdf.

Sullivan, J. 2018. '"Streets Resemble War Zone" – Hangberg Protest, Hout Bay'. *SA People News*. Last accessed 3 July 2018. https://www.sapeople.com/2017/09/12/photos-hangberg-protest-hout-bay/.

Tefre, Ø.S. 2010. 'Persistent Inequalities in Providing Security for People in South Africa – A Comparative Study of the Capacity of Three Communities in Hout Bay to Influence Policing', MA Thesis. Norway: University of Bergen.

Thompson, E.P. 1971. 'The Moral Economy of the English Crowd in the Eighteenth Century', *Past and Present* 50: 76–136.

Union of South Africa. 1950. 'Population Registration Act: No. 30 of 1950'. *SA Government Gazette*.

Von Holdt, K., M. Langa, S. Molapo, N. Mogapi, K. Ngubeni, J. Dlamini and A. Kirsten. 2011. *The Smoke that Calls: Insurgent Citizenship, Collective Violence and the Struggle for a Place in the New South Africa*. Johannesburg: Centre for the Study of Violence and Reconciliation (CSVR).

Wit, E., et al. 2010. 'Genome-Wide Analysis of the Structure of the South African Coloured Population in the Western Cape', *Human Genetics* 128(2): 145–153.

Zahar, M.-J., and E. McCandless. 2020. 'Sustaining Peace One Day at a Time: Inclusion, Transition Crises, and the Resilience of Social Contracts', *Journal of Intervention and Statebuilding* 14(1): 119–138.

Interviews

Fisher A. 2018. Interviewed by Fiona Anciano, 3 July 2018. Hangberg, Cape Town.
Fisher B. 2018. Interviewed by Fiona Anciano, 5 July 2018. Hangberg, Cape Town.
Fisher C. 2016. Interviewed by Conrad Meyer, 24 May 2016. Hangberg, Cape Town.
Fishers D and E. 2016. Interviewed by Fiona Anciano and Conrad Meyer, 20 May 2016. Hangberg, Cape Town.
Fisher F. 2015. Interviewed by Fiona Anciano, 6 March 2015. Hangberg, Cape Town.
Fisher G. 2015. Interviewed by Fiona Anciano, March 2015. Hangberg, Cape Town.
Fisher H. 2016. Interviewed by Conrad Meyer, 24 May 2016. Hangberg, Cape Town.
Poachers. 2016. Focus Group with Poachers conducted by Fiona Anciano and Conrad Meyer, 17 May 2016. Hangberg, Cape Town.
Zille, H. 2017. Premier of the Western Cape. Interviewed by Laurence Poper, 26 September 2017. Cape Town.

5 Spectators of protest

Concerns from an online neighbourhood facebook group

Yusra Price

A protest over the road

It was a weekday in 2018 when I left my home for work a little later than my usual 06:30 routine. The daybreak was orange and the sun was soft. The crisp morning air was already giving way to the afternoon sun. The area in which I live only has one entrance, and as I approached it, I noticed a few cars driving back in. This was unusual to see. I looked up and noticed thick black smoke coming from somewhere close by. As I drew close, I noticed people gathered at the entrance that leads onto Prince George Drive and I was signalled to slow down by an onlooker[1]*; a fellow resident. I rolled down my window and a man approached to say that people were being prevented from leaving because residents of Vrygrond were protesting and stoning cars that attempted to leave. I scanned the residents standing at the entrance of our area who were watching the situation closely; as if the road was a window that separated us from them and allowed us to observe them from a safe distance. I saw a woman in her purple gown with a mug in her hand, taking small sips as her eyes were fixated on the unfolding events not more than twenty metres from where she stood. I turned around and went back home; I was not going to leave my area that morning.*

Whilst the protest unfolded and clashes between police and Vrygrond residents continued, I took to social media to see how people in my area were discussing the protest and protesters alike. I was curious about two things: (i) whether people in my area vocalized concerns about the protest, and (ii) what kinds of perspectives would emerge, especially since we were unable to leave our area due to the single entrance being blocked the whole morning.

Our Marina is a public Facebook group for my neighbourhood and has a membership of approximately two thousand members.[2] This is hardly representative of the whole neighbourhood but is enough to generate a sense of neighbourliness. The Facebook group page is filled with mundane things, from requests for good electricians, to missing animals, to goods and services for sale, homes and rooms to rent and sometimes beautiful pictures of our lovely scenery by an inspired resident. A space for us, as residents of Marina Da Gama, to participate in neighbourliness (Cerwonka 2004). There is conflict from time to time, but there is also always a strong reminder by members and group administrators to maintain the ethics of engagement, which are to engage positively and not destructively.

My personal engagement with the page is minimal and at best superficial, with the exception of my posts that contested the types of narratives that emerged from users about the protest.

When the protests in Vrygrond happened in 2018, the Facebook page turned its attention away from our neighbourly posts to the protest action that prevented our mobility and, for many, posed a threat to our safety. People began to post traffic updates because there were some road closures across our shared traffic light. Many posted or requested updates on the state of the protest out of curiosity or out of concern for leaving the area. The group members were lively and highly responsive. The time elapsed between comments was a few short minutes and sometimes even seconds – instant communication. The way conversations unfolded was disproportionate, with commentary that characterized protesters as indignant, ungrateful, dangerous and entitled, outweighing more sympathetic responses. I was horrified at the prejudiced undertones of comments that described protesters as "stragglers" or "squatters" to refer specifically to migrants or refugees in the area. Using the term "squatters", for example, speaks to people's relationship with legitimate ownership or occupation of land. In South Africa, this is tied to a racist model of displacement that emerged from apartheid, and 'squatting' emerged as a means for people to make a life that works for them.

As a member of Our Marina, I grew concerned with the silence of other users and the absence of counter-narratives. In my discomfort, I thought surely I was not the only observer who was deeply unsettled by the abrasive commentary and the unrestrained willingness to publicly post such comments. It was as if there was no concern of wider public scrutiny and as though their sentiments were shared by other users.

Shah (2013) mentions how the World Wide Web should not be underestimated with its immense capacity to archive. For movements like #BlackLivesMatter and #FeesMustFall, for example, the archiving and productive capacity of the internet helped users maintain the online and offline momentum of these movements. The hashtag for these respective movements became a means of developing "alternative models of democratic self-organization" (Juris 2012: 261) that did not rely on the state to make space for their concerns. Instead, protesters forged their own claimed spaces on the streets, for example, to draw the state's attention (Cornwall 2004, as cited in Mkhize 2015). The hashtag demonstrated the far-reaching and mutually constitutive nature of online and offline engagement (Juris 2012; Shah 2013). To date, there has been growing literature that focuses on protesters, supporters and digital publics; exploring the role of digital activism (Shah 2013); the use of digital spaces and technologies (Daniels 2016); and what can (or cannot) be known about the role that digital publics play in shaping other publics and subjectivities.

Close attention has been focused on the way that digital spaces are used to produce narratives and discursive practices aimed at political change using online spaces (Bonilla and Rosa 2015; Diepeveen 2019). In the case of protest, larger movements have shaped our understanding of citizen action in digital spaces that rely on hypervisibility. In turn, this has diminished our attention to the visibly silent local protests like the one I address here. Our interest in political framings of protest has become keenly oriented towards hashtags, for example, but spaces like Our Marina are somewhat overlooked. Yet, this particular Facebook group (among others like it) became a space in which people represented and expressed their political subjectivities pertaining to protest and, more broadly, their perceptions of democracy. In this chapter, I take a different angle on the same topic, exploring the narratives that emerged in a digital space created for a physical neighbourhood.

The digital online space is considered in this chapter as a space of gatherings and conversations that take place in daily life. In a recent special edition journal article titled 'Rethinking Publics in Africa in a Digital Age', the authors highlight the proliferation of digital spaces and devices in Africa as an opportunity to rethink research interested or located in the 'public' sphere (Srinivasan et al. 2019; Diepeveen 2019). Their work demonstrates that our lives are intertwined with the digital world. Digital spaces are characterized by the speed and proliferation of information and communication with a global reach – they are both a characteristic of and a contributor to globalization (Nyamnjoh 2004). Information circulates widely, and anyone with access to a mobile device and internet service can participate (Srinivasan et al. 2019). Whilst this number is still limited by infrastructure and resources, it is growing. Digital spaces can include websites; social media platforms such as Facebook and Twitter; messenger applications such as WhatsApp or Skype; and virtual gaming realms among others (Eynon et al. 2017). What unites these platforms is that they are embedded rather than distinct from the workings of contemporary society.

This chapter responds to the call by Srinivasan et al. (2019) to explore digital spaces as a means of revisiting 'who' our publics are and 'where' to find them. In doing so, it follows the authors (Srinivasan et al. 2019) in making an explicit shift from perceiving the public as only 'face-to-face' towards less tangible conceptions of 'public'. 'The public' is an evasive category of analysis, characterized by intangible boundaries and realized "in fleeting moments when people engage with circulating information" within a digital context (Srinivasan et al. 2019: 8). People 'meet' in virtual spaces when they converge on shared interests through forums, blogs, live chats or commentary; and these dialogic spaces *represent* and shape their social and political realities (Srinivasan et al. 2019). Within these meeting spaces, people communicate in ways that are complex, dynamic and, most importantly, open-ended (Srinivasan et al. 2019; Brinkman 2019). The impetus is in accordance with the disruptive effects that digital transformation has had on our political debates and mobilizations; the power of the circulation of events and information from official news outlets to anyone with a recording phone; and the "diversities in social organization and ideas of citizenship" (Srinivasan et al. 2019: 3).

I focus specifically on the creation and contestation that surrounds 'publics' in moments of protest. In doing so, this work contributes a new approach to the broader literature on social media and protest. It starts from the premise that engagement in digital platforms can be superficial as well as substantive: people who stay in physical proximity meet online through 'fleeting moments' to engage with circulated information and common interests; but our engagement on digital platforms can also shape and reshape our political subjectivities that manifest in physical-social realities (Srinivasan et al. 2019).

The digital space that I study is my neighbourhood's Facebook group – Our Marina – which was established to serve the materially affluent neighbourhood in which I live. Our neighbourhood is situated opposite its materially deprived counterpart – Vrygrond – and the chapter focuses on a period during 2018 when protesters in Vrygrond were contesting their marginalization. I focus on the narratives within Our Marina that resisted these resistors, the varied responses to these narratives, and the implications they have for continued civil disparity in Cape Town.

I was not only an observer of the conversation but engaged as a resident with the people of my community, often in an attempt to counter the strong, negative opinions that emerged. By unearthing and analysing the complexity of these narratives, we gain important insights into battles over political legitimacy, which in turn provides us with a better understanding of the social contract in post-apartheid South Africa and the diverse values that underlie it. In undertaking this analysis, my chapter complements emerging literature on digital spaces that examines online attempts to close, dismiss or counter-protest, as well as responses to these manoeuvres from digital publics that sit in inconspicuous parts of the digital world.

In the analysis that follows, I demonstrate how, as online contributors to discussions and debates *about* the protests in Vrygrond, Our Marina residents positioned themselves as spectators of protest. In doing so, I develop the concept of the 'spectator citizen', which refers to those who have purposefully *distanced* themselves from the protest and protesters alike through particular notions of citizenship – essentially making themselves third parties to the protest, but remaining fundamentally separate from it. The identity of the spectator citizen is bound up with attempts to narratively demarcate and solidify ideas of what constitutes and represents a 'decent' neighbourhood in the context of a democratic nation (Cerwonka 2004). It illustrates the extent to which layers of separation to race and class divisions in Cape Town are distilled into people's ideas of belonging. The contestation that surrounded attempts to construct and maintain a 'spectator citizenship' highlights the heterogeneity of narratives that exist on the Facebook group, which can in turn be located within a broader social, political and historical landscape. The multiplicity of these narratives reflects the complexities and challenges of forming a young democratic state.

This chapter is organized into two parts. Firstly, it describes the protest and the responses of protesters, the state and Our Marina users. I highlight the way the social contract was derived and used to give support, contest or delegitimize protest. I also take into account the way history is embedded in how people perceive democracy and the historically rooted responses that emerge through specific sites of contestation. Secondly, I will demonstrate how the discursive practices of Our Marina users, as spectator citizens, sat in contention with one another in terms of either dismissing or supporting the protest. A discourse analysis of this space provided insight into the way online platforms, like Facebook, provide space for expressing and reshaping political subjectivity with regards to protest. The chapter concludes with a word on the significance of digital publics and academic inquiry into online spaces.

Separated by concrete and history

Marina Da Gama was established between 1970 and 1980 for the purpose of a new urban-lifestyle estate, demarcated for high-end living for permanent homeowners (Parfect and Power 1997).[3] The area was erected during the forced removals that occurred in Cape Town between the 1950s and the 1980s under the Group Areas Act (Brown 2005). The entire area is demarcated by an extended white wall and sits along Prince George Drive, a road that forms the latter end of the M5 highway as you travel south towards the sea and broadly hosts four lanes with a parallel stretch of palm trees that cut the lanes equally. The white houses in Marina Da Gama are not coincidental – keeping the houses painted white is a condition of living here. We pay a levy and have a Marina Da Gama Association that hosts AGM meetings for community participation in decision-making.[4] There is an explicit construction of community, and to live here, one must subscribe to these conditions. It is similar to a gated community (Lemanski 2004) with one way in and out – except without the gates.

Vrygrond is an informal settlement situated directly opposite the entrance to my section of the Marina – we share a set of robots on Prince George Drive.[5] On my side of the road sits a beautiful suburbia of homes painted white, a gorgeous view of the mountain and a lake for fishing and recreational boating. As you cross the robots into Vrygrond, you travel on an empty road with open field spaces on either side, towards an informal settlement of densely packed homes that sit like a horizon in the distance. The area was established in 1929 and is considered the oldest informal settlement in Cape Town that preceded the advent of apartheid (Brown 2005; Van Heusden and Pointer 2006). The area was 'formally' developed in three phases: first as Sakkies Dorp developed in the 1980s, then Seawinds in the 1990s and finally Vrygrond (meaning 'free land' in Afrikaans) in 1999 (Brown 2005). During apartheid, many people were forcibly relocated to this area – including the former residents of District Six, which had the highest record of sixty thousand people forcefully removed.[6] Since the end of apartheid, these residents have been joined by internal and external migrants, as well as refugees who have established homes in Vrygrond.

The area itself has a range of establishments and services, including a close-by church service; non-profit organizations like The Compass Trust and True North; educational facilities for children like the Butterfly Art Project, iThemba pre-primary school, Jo's school, Vrygrond Computer Lab and the Vrygrond Community Centre; and entrepreneurial services like Frank's Garden Services and a cleaning service. In addition to these local services and establishments, Vrygrond and Marina Da Gama are within walking distance to the Capricorn Shopping Centre that boasts grocery shops, clothing shops, takeaway food outlets, a gym and a hardware store. The range of establishments offers a rich space for entrepreneurship and community innovation.

On the day of the protest, the residents of Vrygrond protested for an entire day and a few days after that as well. The cause of the protests was the demolition of homes illegally erected on a piece of land called Xakabantu (Ntongana 2018). People's homes had already been demolished three times before because the City of Cape Town (hereafter, the City)[7] had previously declared that piece of land as a nature reserve and therefore unfit for housing (Ntongana 2018). One protester said, 'We are sick and tired of empty promises and the City disadvantaging the already disadvantaged people of Vrygrond by breaking down their houses'[8] (Ntongana 2018). The protester's words are reminiscent of the pre-existing condition of poverty and condemnation of the government's slow reaction. He held the City, led by the Democratic Alliance (DA),[9] accountable for perpetuating the conditions they live in as opposed to fulfilling an implicit obligation to allow them access to housing by providing land on which to erect their own homes. Housing has become a pertinent issue in Cape Town and is scrutinized by activist groups like Reclaim the City that highlight concerns of dignity, safety and non-repatriation of homes in the post-apartheid era.

The police managed to push Vrygrond protesters back far enough from the robots so that they were too far to be seen, but the black smoky clouds and police blocking the entry point communicated unrest to drivers passing by. The police responded to protesters with riot gear, rubber bullets and tear gas aimed to disperse them,[10] which is not an unusual series of actions used to silence protests. The police have played a tenuous role in South Africa. Historically, the police manifested apartheid's brutal force of dictatorship and abetted the constitutional infringements on people's human rights. Our transition into democracy did not render a merciful brigade either. The brutal confrontation with #FeesMustFall student protesters brought into question the condition and competence of the police, who are often criticized for inexperienced leadership and an untrained force that has resulted in gross mismanagement and corruption (Mkhize 2015).

A local online news agency called *GroundUp* published an article that stood out to me,[11] citing a Vrygrond protester saying that "We're not asking for a place like Marina Da Gama" (Ntongana 2018).[12] The article highlighted the long-standing separation between Marina Da Gama and Vrygrond, promulgated by apartheid's separatist ideologies. The explicit design of apartheid rested on an underlying premise of 'apart-ness', which saturated every aspect of daily life for people in Cape Town (Lemanski 2004; Brown 2005; Wilkinson 2000). The effects of apartheid ran deep during those forty-six years of active oppression and still continue today through macro- and micro-aggressions, as well as discrimination. The physical remnants of apartheid remain etched into the infrastructural lives of many communities in Cape Town. Emphasis was placed on 'spatial distancing' from white people, who were given central areas of land that had the most beautiful landscapes, views and surroundings (Lemanski 2004). The naturalization of this spatial division carried and facilitated social division (Cerwonka 2004). These social and spatial disparities between South Africans have further been affixed by economic disparity, thus producing the multiple layers of continued separation between people.

The protest in Vrygrond occurred within the context of historical dispossession and displacement, as well as contemporary economic and social inequalities. For protesters, the daily conditions of poverty and the accumulated years of empty promises prevented them from making a home or gaining a sense of belonging as democratic citizens of South Africa. For Our Marina users, the disruption to daily life and impeded mobility centred on the indignant and destructive behaviour of protesters. The criminalization of protest(ers) by the state and other citizens demonstrated the historically rooted framings of 'the other' (Wacquant 2009), specifically in the way protesters from informal settlements have been labelled as chaotic and dangerous by popular media and onlookers.

As protests unfolded, the relationship between residents and the state, represented by the municipality and police, was outlined by a community representative named Michael Khumalo (Ntongana 2018). He mentioned that residents sent an urgent invitation to the Mayor's office and SAPS (South African Police Service) to attend a community meeting and answer people's questions so that the issue would not get out of hand. However, both the police and the City were unavailable, citing previous engagements. Khumalo's statement highlighted a growing frustration among South African citizens with the government making decisions about development, or in this case, destroying housing, without proper consultation (Mkhize 2015). His statement further highlighted the expectations he held of local government to engage with its citizens, given that to do so would characterize democratic governance. The social contract was envisaged by Khumalo as a promise by the state. By remaining silent or refusing to engage with residents on their terms, the City strategically used its silence to dismiss the importance of residents' claims about housing. Thus, the City and protesters did not conceive the social contract in similar terms.

The transition of South Africa from an oppressive state to a democracy can be understood through the notion of a social contract. In South Africa's conception of a democracy, the aim was to redress the one-sided and hierarchical coercion of apartheid's values onto its citizens. There was a perception that the law would play a role in this redress and abet the undoing of formerly oppressive structures. To some extent, the social contract includes a rights-based approach for the state to ensure basic human dignity via the laws of fundamental freedom and entitlement to decent living conditions (Van Rensburg 2013). This particular notion of the social contract assumed the homogenous acceptance of the state's terms by all citizens, as well as the assumption of a top-down form of delivery. The protesters from Vrygrond, however, demonstrated a dimension of the social contract that challenged the state. Protesters highlighted the necessity of interpersonal engagement between citizens and local government in order to meet any basic provisions.

A social contract was constituted not only by the terms of provision, but also on participatory governance that values the input of local residents (Mkhize 2015). Furthermore, residents brought into question the integrity of the state when Khumalo explained:

> We wanted to say to the City, if it is correct that law enforcement should not demolish houses in the absence of an eviction order from the court, why then can we not agree to no more illegal demolition of the houses until the City approaches the court and gets a legal-go-ahead?
>
> (Ntongana 2018)

Here, Khumalo acknowledges the role that the law plays in our current democracy but calls into question the possible breach of that agreement by police and by extension, the state. His concerns further imply the way notions of the social contract have shifted and barely resemble the participatory governance that shaped early ideas of democracy. At least insofar as the social contract is a promise from the state to its citizens, the state has failed to deliver.

The protests and what unfolded between police and protesters were visible insofar as I could witness as I drove by and what was published by local news agencies. But public participation and perception of the protest was largely missing from news reports – although this did not mean that the public was silent. Here, online communicative spaces provided a broader understanding of perceptions of protest. By looking at Our Marina and my own participation to some degree, I gained a sense of how users understood and discussed the protest using particular discourses around citizenship. Although a group like Our Marina is not representative of the entire neighbourhood, the significance of digital publics sits in the unfettered realm of people voicing their opinions and the way these opinions coalesce into discourses that shape engagement and interaction both online and offline. The latter part of this chapter will look at how Our Marina users produced narratives to explain how and why the protest emerged and unfolded. Through a discourse analysis (Dunn and Neuman 2016), I interrogate the extent to which engagement in digital spaces provides us with important insights into the social contract in post-apartheid South Africa, and into the kinds of complexities and challenges posed in the formation of a young democratic state.

Online conversations and perceiving protest

The protests escalated, and late one evening I heard a loud bang followed by several smaller ones. It still unsettles me how familiar I am with these sounds and the conflict they denote. The next morning, Our Marina residents produced pictures of the road-facing complex scorched in brown and videos of burning tyres on the palm tree island. A petrol bomb was hurled at the complex and a few days later, when I drove out of my area, I saw the large brown scorch mark that stained the white wall with an explosive blemish. I returned to the Our Marina Facebook page and scrolled through the commentary on one of the posts.

The popular narrative of the 'dangerous township' emerged quite strongly in Our Marina conversations. The area was often characterized as a 'war zone' and was once likened to Syria. It was as if people's living conditions in Vrygrond were not understood as conditions of poverty and a failure of the state. Rather, discourses of danger (particularly in reference to refugees) and violence ('gang-related') framed users' understandings of people's living conditions and, furthermore, their morality. In this understanding, Vrygrond posed a threat to those who were vulnerable. Paradoxically, the vulnerable were not situated within Vrygrond, but rather outside it. Discourses of danger strongly motivated people's comments, and as Lemanski (2004: 2) argues, contributed to the 'architecture of fear'. This is understood as the cartography of apartheid, underwritten by fear of the other, that today remains in the spatial and temporal distribution of fear of crime that people experience.

The burning of tyres was not so much the issue as the destruction of property, the tampering with robots and the use of petrol bombs. Discourses of criminality were used to condemn protesters and fuelled narratives of fear. Often, conversations included questions of safety, validated fears about the protesters and led to tips being provided on how to protect oneself if protesters dared enter Our Marina. These fears were exacerbated with every post that updated the group about the protest.

In a single thread of commentary, multiple users converged over the protesters' actions and further situated the latter's deviance within their locale. The phrase 'criminal fringe' was used to describe Vrygrond, while one user in Our Marina referred to themselves as 'stoic' – a single word that implied not only a moral contrast to 'these people', but also carried a history of constructing 'the other'. A sense of fear came through strongly when residents described the 'situation' in Vrygrond. In the form of questions, people asked whether the protest 'flared up' again or what 'sparked [the] violence'. There was no acknowledgement of continuity or resurgence, but rather a dismissal of the historical and contextual factors shaping people's realities – on both sides of the road. The protest was understood through notions of fire and with agitation; and with local government not conceding to residents' so-called demands, the situation could potentially flare up again. In the same vein, protests/protesters were positioned as combustible elements that could, with the right catalyst, 'spark violence'. This was an attempt to dehumanize protesters by comparing them to weaponized objects. The protests were further described as a 'horrific situation', 'anarchy', 'pure vandalism' and 'chaos'. Users in Our Marina harnessed language to position protesters outside of the moral framework of good, democratic citizenship; thereby further establishing a clear separation between themselves and protesters. In doing so, they made room to watch and discuss the protest without having to claim any kind of involvement beyond being victims of potential violence.

A few other online members and I managed to coax some debate and draw attention to police brutality or the ineptitude of law enforcement to deal peacefully with protesters. I typed a long post that detailed my concerns with the way protesters were being portrayed. Some users commented to support my concerns, while some used the 'like' function. A few likes on a few of my comments and posts were akin to quiet nods in agreement – as if doing so did not attract much attention but still communicated agreement in our debates. Creating dialogue was emotionally exhausting, especially in the face of persistent members who did not let go of their reservations. We focused on the long-term effects of apartheid on people's realities and tried to blur the lines of separation and diminish spectatorship. We urged users to reconsider the ways that our neighbourhood, by remaining silent and in doing so being complicit, perpetuates the inequality experienced by our neighbours. The discourse we drew on bore a strong resemblance to the way silence was read in the #BlackLivesMatter movement. However, the silence we were referring to was the extent to which our engagement online did not seem to translate to offline mobilization or support towards the protest. Instead, in taking footage of the burning tyres at night and attracting negative conversation online, it seems as though Our Marina users were contributing to the silencing of protesters.

The way online conversations unfolded revealed the tensions of competing discourses in Our Marina and brought into sharp focus people's understandings of contemporary South Africa, specifically, the way the social contract is envisaged from different perspectives. Cerwonka (2004: 4) mentions 'spatial practices' that she defines as 'cultural practice[s] by which people construct or define place'. Here, spatial practices can include the less obvious, non-physical practices that underpin the conditions of belonging, such as interactions on the Our Marina Facebook group or sharing a path for walking along the lake. Cerwonka (2004: 4) adds that spatial practices further refer to 'social practices that function to define space and how they are embedded in a wide range of power relations'. The contrast between Marina Da Gama and Vrygrond, as highlighted by one protester, demonstrates the way historical spatial practices have generated certain social and political orders (Cerwonka 2004), which was exemplified by interactions between protesters, local government and police. Furthermore, these practices were also revealed by the way users in Our Marina assumed the position of spectatorship.

In a post-apartheid South Africa, I could argue that there is a need for people to reconstitute their connections to a place, since historically this was predetermined for them and even [racially] naturalized in the cartographies of apartheid (Cerwonka 2004; Lemanski 2004). I am reminded of Benedict Andersen (1991, as cited in Cerwonka 2004), who focuses on state practices to locate his notion of an imagined community (nationality). According to Andersen (1991), nations are constituted through an imagined organic connection between people, culture and place. Beyond apartheid, 'organic connections' have been forged within cartographies of separation and continue to manifest material disparity and unequal access to social goods. Community-making has become a means for people to forge connections, yet these remain difficult to link across class, racial and financial boundaries, especially when community-making is necessitated by different needs or aspirations.

Our Marina users were aware that land was an issue, but many argued that Vrygrond residents could not rely on the government for 'hand-outs' and implied further that only hard work and determination would get them out of their living situation. This made their criminality more plausible and Our Marina residents' fears more valid. What was interesting, however, was the call for the police to 'protect' citizens. This invoked an understanding of a social contract wherein the police, as representatives of the state tasked with maintaining order, were called upon to uphold their responsibility towards citizens. The question arises, however: who comprises citizens? And what comprises law and order? At this point there was a clear sense, from the point of view of Our Marina commentaries, that within our current context there are forms of discipline for those who do not live within the parameters of the 'nation-state' (Cerwonka 2004). In other words, if protesters continued to disrupt and cause damage to property, then they forego any entitlement to state protection and must be met with the law. For some users in Our Marina, this was grounds for dismissal of the protest entirely. Users showed clear abhorrence for protest(ers) and resistance to any alternative ways of understanding the protest, as if the prejudice they showed was a justified expression of rejection of non-democratic and non-compliant people. There was a clear interplay between being a spectator citizen and using notions of discipline and morality to underlie their ideas of what the social contract actually denotes.

Cerwonka (2004) reminds us that ethnographic data should be read in relation to history in order to show how criminality and moral behaviour are important for forming collective identities – such as the identity of the 'Rainbow Nation'. In order to conceptualize the workings of the social contract, history is seen as both relational and relevant. The place of history is contextual, and for South Africa, in a post-apartheid milieu, history is embedded in the present workings of how people perceive democracy. More specifically, history is not only embedded but internalized in South Africa's spatial, social and political landscape. When residents feared harm, they did not refer exclusively to physical harm, but harm to their imagined communities. By historically situating protesters and Our Marina users' ideas of the state and what underscored the social contract for each of them, we are made aware of how the social contract is not a homogenous set of precepts that shape our democracy. Instead, I note that the heterogeneity that composes the state, protesters and Our Marina users depicts multiple perceptions of what the social contract means and can mean depending on where we are situated – within history and contemporary disparity.

As many have discussed around digital activism, the way online interactions have manifested as aggregates of people for protest has drawn significant attention, specifically, the aggregation of people in support of protest. However, in the case of Marina Da Gama, residents did not gather for the sake of support, but for the sake of spectatorship. I used the terms 'spectator citizens' to describe the ways users separated themselves from protest, showing that the only time they gathered around protest was to observe and not participate. To make sense of how Our Marina is situated within the broader discussion of digital publics, I argue that it is a digital public space that people use to express or reshape their political and social lives (Srinivasan et al. 2019). In the same way that the neighbourhood page was used to discuss the indignation of protesters, it also became a space for dialogic engagement, as I and others intervened to give insight into the role of contestation when it comes to protest in South Africa. Although not representative of the entire neighbourhood's values, the political subjectivities that emerged on Our Marina provided insight into the ways the social contract in a democratic context was shifted and composed by protesters, the state and some users in Our Marina.

Contrasting these neighbourhoods points to how different meanings of what is right and wrong derive from people's living conditions (Siméant 2015). In these instances, citizenship and the social contract are positioned and located differently. Firstly, for the few protesters who spoke to news reporters, their sense of citizenship and belonging is located in their relationship to the state. They emphasized the developmental role of the state to alleviate their struggles and the dialogic state to consult with its residents – the failure to do so put Vrygrond residents in a precarious place (Von Schnitzler 2014). From the perspective of the state, the responsibility to uphold the values of democracy was communicated through the actions of the police. The emphasis on law and order and a failure of

Vrygrond protesters to abide with this resulted in their criminalization and penalization. From the perspective of Our Marina residents, based on the narratives that emerged, there was a strong emphasis on the security role of the state to protect what already existed and a leveraging of the social contract to repress protesters. For some, on the other hand, the responsibility slanted more towards the state with some responsibility placed on the protesters as well. Overall, the locus of responsibility bounced between the state and protesters, further illustrating the disparity of values in our current democracy.

By locating my empirical data in the virtual realms of our sociality, I was able to include more spaces in which the 'public' resides, specifically, a group of people who occupy a demarcated living space both offline and online. This shows how contentious understandings of democracy emerged, and some of the problematic shared narratives would not have emerged as easily outside of the Facebook group. I was able to shift the focus from the citizen-state relationship to include spectator citizens who, in Our Marina, developed their own sense of what was moral and just with regards to the protest. As a result of this expansion and shift of ethnographic inquiry to online spaces, I was able to gain deeper insight into the different ways the social contract was derived from particular social, political, historical and economic standpoints – which in turn shaped the way the protest was perceived.

Access to social media by specific groups affected by protest enabled them to shape the meaning of protest(ers). Insight into online spaces such as Our Marina brings into question the ability of such spaces to incite change that has an effect on our offline lives. Perhaps not to the same degree as larger movements like #BlackLivesMatter, but in the subtle interactions we have with and in relation to one another. Interrupting the prejudiced discourses that were being repeatedly reproduced in Our Marina actually opened up the platform to more nuanced discussions. It is evident that gathering on social media presents us with opportunities for engagement that are a significant marker of contemporary socializing. There is a need for greater insight into the ways we might claim that online and offline engagements are mutually constitutive and, if so, how we make sense of the less spectacular forms of influence.

Notes

1 Prince George Drive forms the latter part of the M5 highway south that leads into the Peninsula area in which Muizenberg is situated.
2 Our Marina is a pseudonym in place of the public Facebook page. The ethical conduct of researchers in a digital space is still considered new and according to Pauwels (2006), researchers have to develop norms for themselves. The group remains active, so I chose to anonymize it.
3 The vision for the construction of the area was outlined in a pamphlet – g 'Marina Da Gama – Cape Town', Patrick Robert Brian Lewis Papers, 1939–1987, Item Nr C475.
4 http://www.mdga.co.za/ is the official site.
5 Robot is South African English for a traffic light.
6 In this survey, Brown (2005) includes short anecdotes from Vrygrond residents who lament their struggles with housing. The demographics of the area is largely

'coloured' and 'black', and the ages range from very young children to the elderly who were reaching 80 years old at the time.

7 Referring to the municipality.

8 https://www.groundup.org.za/article/we-not-asking-place-marina-da-gama-says-vrygrond-protester/.

9 In South Africa, provincial elections determine which political group will govern a particular province and city. In Cape Town, the DA has been a popular choice since 2009 and has won elections ever since.

10 Footage of this can be seen at https://www.youtube.com/watch?v=fWIrCkBCRVE.

11 https://www.groundup.org.za/about/ – GroundUp puts forward a human-centred approach that emphasizes human rights-based journalism, and often their articles provide the perspective of those they seek to write about.

12 https://www.groundup.org.za/article/we-not-asking-place-marina-da-gama-says-vrygrond-protester/.

References

Andersen, B. 1991. *Imagined Communities: Reflections on the Origin and Spread of Nationalism*. New York: Verso.

Bonilla, Y., and J. Rosa. 2015. '#Ferguson: Digital Protest, Hashtag Ethnography, and the Racial Politics of Social Media in the United States', *American Ethnologist* 42(1): 4–17.

Brinkman, I. 2019. 'Social Diary and News Production: Authorship and Readership in Social Media during Kenya's 2007 Elections', *Journal of Eastern African Studies* 13(1): 72–89.

Brown, F. 2005. 'Housing Crisis in Cape Town, Western Cape, 1994–2004', in *From the Depths of Poverty: Community Survival in Post-Apartheid South Africa*. Durban: Centre for Civil Society, pp. 83–107.

Cerwonka, A. 2004. *Native to the Nation: Disciplining Landscapes and Bodies in Australia*. Minnesota: University of Minnesota Press.

Cornwall, A. 2004. 'Reflections on Issues of Power and Difference in Participation in Development', in S. Hickey and G. Mohan (eds.), *Participation from Tyranny to Transformation: Exploring New Approaches to Participation in Development*, London: Zed Books.

Daniels, G. 2016. 'Scrutinizing Hashtag Activism in the #MustFall Protests in South Africa in 2015', in B. Mutsvairo (ed.), *Digital Activism in the Social Media Era: Critical Reflections on Emerging Trends in Sub-Saharan Africa*. Switzerland: Palgrave Macmillan, pp. 175–193.

Diepeveen, S. 2019. 'The Limits of Publicity: Facebook and Transformations of a Public Realm in Mombasa, Kenya', *Journal of Eastern African Studies* 13(1): 158–174.

Dunn, K.C., and I.B. Neumann. 2016. *Undertaking Discourse Analysis for Social Research*. Michigan: University of Michigan Press.

Eynon, R., J. Fry and R. Schroeder. 2017. 'The Ethics of Online Research', in *The SAGE Handbook of Online Research Methods* (Vol. 2), pp. 19–37.

Juris, J.S. 2012. 'Reflections on #Occupy Everywhere: Social Media, Public Space, and Emerging Logics of Aggregation', *American Ethnologist* 39(2): 259–279.

Lemanski, C. 2004. 'A New Apartheid? The Spatial Implications of Fear of Crime in Cape Town, South Africa', *Environment and Urbanization* 16(2): 101–112.

Mkhize, M.C. 2015. 'Is South Africa's 20 Years of Democracy in Crisis? Examining the Impact of Unrest Incidents in Local Protests in the Post-Apartheid South Africa', *African Security Review* 24(2): 190–206.

Ntongana, T. 2018. '"We're Not Asking for a Place Like Marina Da Gama" Says Vrygrond Protester'. *GroundUp*. Last accessed 15 May 2019. https://www.groundup.org.za/article/we-not-asking-place-marina-da-gama-says-vrygrond-protester/.

Nyamnjoh, F.B. 2004. 'Globalisation, Boundaries and Livelihoods: Perspectives on Africa', *Identity, Culture and Politics* 5(1–2): 37–59.

Parfect, M., and G. Power. 1997. *Planning for Urban Quality: Urban Design in Towns and Cities*. London: Routledge.

Pauwels, L. 2006. 'Ethical Issues of Online (Visual) Research', *Visual Anthropology* 19(3–4): 365–369.

Shah, N. 2013. 'Citizen Action in the Time of the Network', *Development and Change* 44(3): 665–681.

Siméant, J. 2015. 'Three Bodies of Moral Economy: The Diffusion of a Concept', *Journal of Global Ethics* 11(2): 163–175.

Srinivasan, S., S. Diepeveen and G. Karekwaivanane. 2019. 'Rethinking Publics in Africa in a Digital Age', *Journal of Eastern African Studies* 13(1): 2–17.

Van Heusden, P., and R. Pointer. 2006. 'Subjectivity, Politics and Neoliberalism in Post-Apartheid Cape Town', *Journal of Asian and African Studies* 41(1–2): 95–121.

Van Rensburg, L.J. 2013. 'A Human Rights-Based Approach to Poverty: The South African Experience', in N. Kakwani and J. Silber (eds.), *The Many Dimensions of Poverty*. London: Palgrave Macmillan, pp. 165–183.

Von Schnitzler, A. 2014. 'Performing Dignity: Human Rights, Citizenship, and the Techno-Politics of Law in South Africa', *American Ethnologist* 41(2): 336–350.

Wacquant, L. 2009. *Punishing the Poor: The Neoliberal Government of Social Insecurity*. Durham: Duke University Press.

Wilkinson, P. 2000. 'City Profile: Cape Town', *Cities* 17(3): 195–205.

6 'This is our water!'

The politics of locality and the commons in the city of Bulawayo

Mmeli Dube and Katharina Schramm

Introduction

Bulawayo's water problems are as old as the city's existence. Due to its geographical location, the city has experienced incessant droughts and water shortages for over a century (Baker 2012; Gwebu 2002: 420–1; Musemwa 2006, 2010). This situation has been characterized by rationing schemes that have limited residents' access to water. Early on, struggles over water contributed to a high degree of political organization among the black residents of Bulawayo who claimed equal access to resources and fought for equal rights of citizenship (Ranger 2007: 163). The resulting political associations played a pivotal role in the liberation struggle.[1] After independence in 1980, the city became a stronghold for opposition parties: first for the Patriotic Front-Zimbabwe African People's Union (PF-ZAPU), and later, from 2000 onward, the Movement for Democratic Change (MDC). Ever since, the management and ownership of water remained both a source of conflict and a political asset for the state, the local council and citizens alike.

This chapter traces two distinct moments of contestation over water, in which different understandings of water as a common good surfaced and conflicting expectations about the respective roles of citizens and authorities emerged at the level of the state as well as the city council. We work with the concept of the moral economy in order to pay close attention to the dynamics between the articulation of political values and the role of affective mobilizations in political protest. In fact, these two cases illuminate the importance of *locality* in understanding how the legitimacy of the post-colonial state is called into question when demands for social services intersect with conflicts over political recognition and participation. Who should be included and have access to the commons? Who should control these resources, and who can be held accountable when service provision fails? These questions point to specific articulations of moral economies in the plural, thereby demonstrating that positionalities in protest actions are not fixed but emerge through a complex network of relations.

'This is our water!' was a rallying cry in both protests, but its points of reference and its addressees were different in the two cases. In this chapter, we discuss how activists, politicians and citizens mobilized the notion of water as a common good to different ends. Hereby, we will critically engage the concept of the moral economy as a framework of mutual expectations, obligations and relational politics. In reference to the seminal works of E.P. Thompson (1971) and James Scott (1976), Siméant notes the ongoing analytical force of the concept:

> 'Moral economy' implies a relational conception of legitimacy, therefore, one can think about contestation and paternalism together, as one can with patronage and rumor, deference to authority, and the obligation that leaders have to act as 'good authorities'. The concept avoids economic reductionism, challenging the idea of a market or an economy autonomous from the social world, and identifying popular conceptions of right and wrong in economic matters.
>
> (Siméant 2015: 165)

Thompson and Scott situate their discussions in specific historical moments and with reference to the economic and political transformations and shifting market relations accompanying them. Thompson's (1971) analysis focuses on eighteenth-century food riots in light of changing forms of patronage in England; while Scott (1976) deals with 'subsistence ethics', peasant politics and rebellion in colonial Burma and Vietnam. To both scholars, the rage and indignation underlying these historical protests did not arise from a situational rebellion of the belly. Rather, these affective mobilizations were deeply embedded in the social fabric that shaped the normative horizon of the poor. What these authors regard as a larger moral economy was linked to a sense of reciprocity and economic justice that set out obligations for those in power to guarantee a reasonable life for all.[2] When this equilibrium was threatened, people resisted.

The popular protests around water in Bulawayo showed some elements of this dynamic. In both instances, participants called upon national and local authorities to fulfil their obligation, namely to provide equal access to water – a scarce and precious resource. However, both protests and the claims they articulated were not isolated but deeply entwined with broader politics, collective memories and the everyday experiences of Bulawayo residents. Thinking along a wider scope of the moral economy in relation to the common good allows us to consider these protests as forms of 'political subjectivization' (Fassin 2009: para 8) in multiple ways. First, the moral economy framework draws our attention to the sense of expectation and reciprocity that is often associated with service provision. Water provision, just like health, housing and education, is a central obligation for the government – a yardstick by which it can be judged. As a means of subsistence, it appears to stand outside the standards of market indicators. Second, the moral economy puts an emphasis on the profoundly political nature of subsistence rights and struggles over the common good. Integrating the dimensions of the production and circulation of services (economy) with the constitution and use of norms and obligations (moral), the moral economy opens up the space of the political to include matters of recognition and affect in political subject formation (Fassin 2009: para 13). This leads to a third point, namely the articulation of political subjectivity in the very act of protest. While for Thompson (1971), this might have related to the formation of a new class consciousness (Ibid.: para 8), the Bulawayo water protests brought forth simultaneous claims to political autonomy *and* citizenship. In other words, water (and water management) became a moving target, changing its significance according to the body of authority that was being addressed in a particular constellation. In order to draw out the nuances of these different relations and associations, we focus ethnographically on the ways in which people linked their claims to significant points of reference in their historical and everyday experiences. Hereby, we hope to show how a strong sense of locality and ownership could be in line, but also in tension with, understandings of water as a common good.

We approach these questions from different positionalities and disciplinary angles. Mmeli, who is from Bulawayo, actively (and prominently) participated in the 2013–2015 protest action. As an emerging political science scholar and activist, he is writing his PhD on local associations, democracy and citizenship in Cape Town and Bulawayo. Katharina, who is an outsider to Bulawayo and Zimbabwean politics, adds a long-standing research interest in the anthropology of citizenship and political subjectivity. We hope that our discussion will add to broader concerns around the rethinking of political values and practices of resistance, as gathered in this edited volume.

Our chapter will proceed in three stages. First, we will add a historical perspective to the recent protests, introducing Bulawayo's unique position in the national landscape of Zimbabwe. We will then turn to the two protests and give a detailed account of the ways in which the claim 'This is our water!' was expressed. In conclusion, we will draw the two cases together and discuss their broader implications for theorizing the multiple and shifting contestations of moral economies in relation to the articulation of political subjectivity.

Bulawayo: The dry city of kings

Bulawayo is located in the south-western part of Zimbabwe, in a semi-arid region with erratic rainfall and that is prone to droughts (Baker 2012; Gwebu 2002). Nicknamed the 'City of Kings', the city's name, Bulawayo, was derived from that of the royal throne of the last King of the Ndebele, Lobengula, who settled there with his people in the 1870s up to the early 1890s. It was established as a town in 1894, after the settlement was captured by Cecil John Rhodes's British South African Company (BSAC) soldiers during the First Matabele War (Ranger 2007). The Ndebele people were driven away to the surrounding areas by the colonists, but they still considered the city as the fortress of their identity and culture.

Historically, the city's water, roads and other infrastructure – though constructed by black labour – were designed to cater exclusively to the white settler population. In his social history of Bulawayo, Terence Ranger (2007) discusses how white residents (labourers, artisans and capital owners) influenced the city council accordingly. As citizens and ratepayers, they also helped out with loans to finance the expansion of services, including water. However, they insisted that they should 'not be expected to pay for African housing or amenities' (Ranger 2007: 35). Nor should African workers have the same rights as their white compatriots: 'Bulawayo's whites were determined that Africans should have no chance to prove themselves to be enterprising townsmen' (Ibid.: 36; also see Bolding et al. 2000). In fact, citizenship as a means of political participation was limited to whites; blacks remained outsiders to the colonial state (Mamdani 1996).

In the 1920s, despite their privileged access to water, the city's ratepayers complained about water shortages to the Bulawayo City Council (BCC). From the mid-1890s onwards, the city's water provision, power generation and management of lighting had been in the hands of a private company, Bulawayo Waterworks Company (BWC) (Musemwa 2008). In 1924, after years of political pressure from the white settler residents and failed attempts to buy the water and lighting back from the BWC, the city council finally took over the management of water supplies by force. After the expropriation of water management duties, from the late 1920s until 1976, the BCC 'built a major water supply scheme every eight years to cope with its ever-increasing domestic, institutional and industrial water requirements' (Gwebu 2002: 422).

In 1976, the Rhodesian government installed its first Water Act that transferred the powers relating to the construction of dams and the tapping of water from sources such as aquifers from local authorities to the central state (Musemwa 2008).[3] This move compromised the city council's ability to be fully responsible for the city's water supplies. At independence in 1980, the new government inherited the 1976 Water Act. During the early years of independence, the national government faced the immense task of improving the levels of social development for all citizens, particularly in sectors such as education and health. Part of its measures involved several amendments to the 1976 Water Act in 1986, 1989, and 1996 respectively. Those changes, however, did not eradicate the inequalities in access to water (Youde 2010), and water continued to be highly contested in Bulawayo (Musemwa 2006). Despite severe droughts, not a single new dam was built between 1976 and 1992 (Ibid.: 896).

New socio-political challenges exacerbated the already prevailing water problems, resulting in the failure of supply to meet the rapidly rising demand for water in Bulawayo (Gwebu 2002; Musemwa 2006). The promise of independence in 1980 was to transform society through the democratization of access to social services such as water, health and education (Mate 2018). Despite this apparent social contract, in the years following independence, conflicts between the two liberation movements – the ruling Zimbabwe African National Union (ZANU PF) and the Patriotic Front-Zimbabwe African People's Union (PF-ZAPU) – negatively affected the local and national governments' response to the water crisis. The ensuing conflict between the two parties was severe. Between 1983 and 1987, the ZANU PF government launched a programme aimed at rooting out what was said to be dissidents associated with PF-ZAPU. This so-called Gukurahundi initiative, which in Shona translates as 'the early rain which washes away the chaff before the spring rains', resulted in thousands being massacred and more people fleeing from the rural Matabeleland and Midlands regions, and into urban areas such as Bulawayo (Catholic Commission for Justice and Peace 1997; Mpofu 2010).[4] At the height of the conflict, all development projects were suspended as the affected areas were sealed off (Agere 1987, in Mate 2018: 2). The conflict between the city and the state also resulted in the central government withholding funds for developing water infrastructure (Gwebu 2002), and 'hijacking' some locally driven water projects (Musemwa 2006: 246).

In the same era, Bulawayo also saw rapid urbanization due to massive rural-urban migration. The resulting population growth undoubtedly put an additional strain on the city's infrastructure. The local government thus floundered under the burden of expectations regarding improved access to water and other social services (Gwebu 2002). Since the late 1920s, the city has relied on dams located between fifteen and forty-five kilometres around it. Most of these dams were constructed between 1928 and 1975, before the 1976 legislative changes. Moreover, some of its water supply was obtained from groundwater that was pumped from the Nyamandlovu aquifer located sixty kilometres northwest of the city. In addition to the problem of distance, the aquifer, which is still in place, is constantly under threat of over-exploitation and contamination from fertilizers and pesticides by agricultural activities around the area of its location (Rusinga and Taigbenu 2005).

Alongside these infrastructural constraints, party politics have also played out in conflicts over water. Since the country's independence in 1980, the residents of Bulawayo have had a well-documented history of voting for the opposition. In 1985, after the ruling ZANU PF's failed attempt to win electoral seats in Bulawayo and most parts of Matabeleland, the then Prime Minister Robert Mugabe[5] warned the people in those areas against voting for PF-ZAPU. In 2000, when ZANU PF lost the city to the Movement for Democratic Change (MDC), Vice President Joseph Msika labelled Bulawayo residents *imigodoyi* (useless dogs) for supporting the opposition (Mpofu 2010). In 2013, after failing to win parliamentary and local government elections in Bulawayo (and other urban centres), then President Mugabe once again told the Bulawayo residents to 'go get services' from the MDC since they had rejected his party and his government (Majaka 2013). The striking thing about these remarks is the implied connection between service delivery and party patronage. The ZANU PF politicians recognized the obligation of the government to care for its citizens – but the reciprocity would be complete only if people gave full electoral support to them as the party that managed the distribution of resources.[6] They differentiated between deserving and undeserving citizens, and thus linked citizenship rights to unquestioned support for those in power. This reaction further alienated Bulawayo citizens from the national government and consequently exacerbated political tensions.

When ZANU PF nearly lost the parliamentary elections in 2000 and registered huge losses in major cities and towns to the newly formed MDC, the water infrastructure of Bulawayo and other cities became a potential tool for regaining some urban control for the ruling national government. Applying what Resnick (2014: 4) has called 'strategies of subversion', ZANU PF used water as an asset to weaken the opposition. Some of its key strategies included (i) issuing directives that usurped the local authorities' powers in the areas of budgeting, human resources and others, and (ii) withholding funds from the BCC. This, for instance, undermined Bulawayo and other cities' capacity to acquire chemicals for water treatment and spare parts to repair infrastructure (Musemwa 2010; Youde 2010). Amidst the economic meltdown, the funds that had previously been allocated for water and sanitation were now redirected towards military and security forces loyal to Mugabe and ZANU PF (Bratton 2009, in Youde 2010).

Such forms of governance, where social goods are sacrificed for regime security, have put a huge strain on relations between the local and the national governments. Residents of Bulawayo and other affected cities turned their backs on the national government – and opposition against its policies and wider politics grew. In Bulawayo, this also had an impact on the perception of water as a common good. On the one hand, the neglect and apparent active sabotage of the city's water infrastructure constituted a breach of the contract of citizenship and of the national government's role as a social service provider. On the other hand, residents of the affected cities became even more suspicious of that very government. Water management should remain in local hands, because it seemed clear that only the city council would adequately care for the residents and thus be in a position to ensure the fair distribution of water to all. In moral economy terms, what was at stake was not only the social service itself, but also the very authority to whom demands should be rightfully addressed, thereby calling into question the legitimacy of the post-independence state.

Struggling over water and political autonomy: The protests of 2005–2007

The local custodianship over water was threatened when, in 2007, the ZANU PF-led national government issued a directive to the Zimbabwe National Water Authority (ZINWA) to take over water management duties and infrastructure in the big cities, including Harare and Bulawayo, all under the pretext of improving water supplies (Youde 2010). The government cited local authorities' failure to provide adequate water to urban dwellers, which made it necessary for the central government to intervene. The move led to different interpretations and reactions in different localities. Whereas Harare's water resources were taken over almost immediately, Bulawayo put up a fight and warded off ZINWA's advances (Musemwa 2008, 2010; Youde 2010).

'Hands off – this is our water!': Asserting local control of the commons

Under the rallying call 'This is *our* water!' activist groups mobilized the city's residents to articulate their political voice, protest against the government's decision and retain local governance of water. The MDC-Mayor of the City of Bulawayo, Japhet Ndabeni-Ncube, together with councillors and senior technocrats, led the charges against the national government. They convened meetings in volunteers' houses in different suburbs, urging the residents to reject the takeover and stating the BCC's position on the matter. They held meetings with the media, civil society leaders and senior politicians, sharing the implications of the takeover and requesting the support of the council in rejecting it. Human rights organizations and the business community, as well as senior ZANU PF party leaders in Bulawayo and Matabeleland, were also against the ZINWA takeover.

The Mayor and his allies invoked both political and technical arguments against the takeover. Concerning technicalities, the critics articulated serious doubts about ZINWA's capacity to operate professionally and effectively, citing the problem of revenue collection among other things. These concerns were mixed with matters of ownership, identity and belonging. In one of the key interviews to explain the reasons for his council's rejection of the central government's policy, Ndabeni-Ncube employed the notion of water as a common good, together with strong references to locality: 'The people are simply saying this is our baby. The City of Bulawayo is ours. The infrastructure is ours. The infrastructure [that] ZINWA wants to take over is ours. We need our revenue!' (Rukuni 2007).[7] This sense of ownership prioritized the city over the nation.

While the national government was certainly expected to provide the funds to expand and maintain the necessary infrastructure (e.g. the dams), the responsibility was placed on the citizens themselves to manage its resources. The Mayor's claim appeared as the defence of a birthright (Musemwa 2008). He evoked the spirit with which Bulawayo had fought off a private company in the late 1920s and how the city had ever since remained in charge of its resources. This historical analogy is significant, as it prioritizes local autonomy over colonial power hierarchies. After all, the Bulawayo City Council at the time had been solely in the hands of white settlers, and black people had hardly benefited from the city's victory against private control. The fact that the conflict nevertheless played out in the fight against the ZINWA takeover shows the depth of the cleavages between the national and local governments.

Linked to the Mayor's implied reference to the investment that the citizens had made into the city's water infrastructure for over a hundred years was his charge that the national government was trying to destroy the city (Nyoni 2007). Residents did not trust the national government to really care for them or their needs. Given the strongly worded offences that prominent government officials had voiced against them, they had a reason to be doubtful about this. Prominent water activists such as Arnold Payne also led the charge against the national government.[8]

In addition, the Mayor and senior local government officials cited the state agency's failure to provide water elsewhere.[9] Finally, he pointed out the indignity that the people had already been subjected to – unable to bathe or flush their toilets, defecating in open spaces and thereby increasing the risk of disease outbreaks – thus the water crisis was already severe (Ibid.). According to critics, people feared that a takeover by ZINWA was going to exacerbate such challenges.[10] In asserting the city's autonomy and linking it to the affirmation of people's dignity, the Mayor and his allies made a triple reference to the moral economy as a form of political subjectivation. First, he confirmed that adequate service provision was a moral *obligation* for the government in power. Yet, he also insisted that this obligation could be fulfilled only by local authorities (i.e. his own administration), since it was built on intimate knowledge of the people's plight and an ensuing relation of trust. In order to achieve this reciprocity, the government needed a face that could also be held accountable – such as the BCC. Second, he emphasized the need for *recognition* of the subject matter at hand (water and sanitation), as well as of the political subjects who were affected by the problem as city dwellers and citizens. Finally, his invocation of people's indignity rhetorically marked a *transgression* of what could be considered bearable – and therefore justified protests on the grounds of a moral economy (see Scott 1976). Interestingly, it was the representatives of the local state who laid claim to the position as the legitimate authority who knew the needs of the people and thus would be able to adequately care for them. The moral economy is thus not a framework of fixed subject positions of provider and recipient. As the case of the Bulawayo water protest shows, these positions are relationally enacted and often fiercely contested.

Awakening residents' collective power

The council-led protests received support from civil society organizations working on human rights and governance issues. These bodies created spaces for political engagement between different stakeholders. Meetings convened by Bulawayo Agenda, for instance, brought civil society leaders, religious authorities, women and youth together to engage in debates on the matter (Mabiza 2013). In public meetings (held in community halls and other spaces), residents publicly expressed strong emotions about the policy. Organizations such as Radio Dialogue conducted intensive public awareness campaigns by distributing audiovisual materials, conducting roadshows and educating the public in their programs. According to one of the civic leaders of the protests, 'almost everyone was united on the need to protect the city from further asset-stripping. We knew that it was not about improving water management but stripping the city of its assets as it had happened [before]'.[11] Other groups like Ibhetshu LikaZulu resorted to the 'disruption of meetings, and using whatever space was available to defend the city'. Collective memory was invoked as well. In particular, people made strong reference to the past 'transgressions' of the national government and argued that it could not be trusted, especially with the management of a precious social good such as water. For example, during one of the meetings held in the city between journalists and the Minister in charge of water at the time, Munacho Mutezo, the Gukurahundi massacre of the early 1980s stood out as a point of reference. Qhubekani Dube, the former leader of Ibhetshu LikaZulu, once asked the Minister, '*Njengoba lake lasibulala nge-Gukurahundi, pho alisoze lisifakele i-poison emanzini yini?*' ('Since you once killed us through Gukurahundi, will you not poison our water?').[12]

The oppositional MDC – through the media, pronouncements in their own meetings and civil society-convened platforms – dismissed the proposed takeover of the city's water reticulation as politically motivated, unjust and unsound (for instance, see Sibanda 2003). To the MDC, some civil society organizations and some scholars, the move by the national government was seen as an escalation of the 'strategies of subversion' (Resnick 2014: 4) and tactics for 're-urbanization' (Kamete 2006: 271) by the ZANU PF government.[13] Quite surprisingly, the issue of the water takeover mobilized people who were previously polarized (or are still polarized on other, mostly national, issues) across political parties. Thus, some senior ZANU PF officials (within the party's national and provincial organs) from the city of Bulawayo and the Matabeleland regions – including former Minister of Home Affairs and then Chairperson of the Matabeleland Zambezi Water Trust, Dumiso Dabengwa, and then Minister of Information, Sikhanyiso Ndlovu – all rallied behind the BCC's rejection of the directive (BPRA 2015; Musemwa 2008). In an unusual feat, they broke ranks with their national leadership and took positions against the takeover of the city's water. When it came to the issue of ZINWA's takeover of water management, local identification trumped political party affiliations. We argue that one of the reasons for the unusual constellation of forces was the shared history of victimhood. Most of the senior ZANU PF officials who stood up against the national government were former members of PF-ZAPU who had joined ZANU PF in the 1987 Unity Accord. They had experienced Gukurahundi and witnessed the previous neglect of the city's infrastructure. Now they feared that their city was once again under siege. Another factor for their stance was the impending 2008 elections, which they feared losing to the MDC again, this time as punishment for the ruling party-led government's policy directive.

Water, authority and legitimacy

In 2008, in what appeared to be a calculated-but-too-late move, ZINWA suspended the takeover of water in Bulawayo. The party in government, ZANU PF, feared an electoral backlash (Nyathi 2008). Despite this concession to the protesters, the party lost both the local parliamentary elections across the city as well as the first round of the national presidential elections. The fact that voluntary organizations known to be aligned with the ruling party, ZANU PF – such as the Bulawayo United Residents Association (BURA) – had also joined in the support for the BCC to retain control of its water (Mabiza 2013) clearly illustrated how emotionally and politically charged the water issue was. It also demonstrated a shift in loyalty that fits with Scott's (1976) observations that in the moral economy framework, a leader's legitimacy depends on their 'ability to provide those necessities of life' (Scott 1976, in Siméant 2015: 164). The government had presented the ZINWA takeover precisely as a move to improve the water situation, i.e. to fulfil its obligations and to provide the citizens with water – a necessity of life. Ironically, the move had the opposite effect, and it appeared as if the government was stealing the water from its citizens. Notions of water as a common good, sanitation as a social service and revenue as a means of redistribution became inextricably interlinked and highly contested. Misjudging the significance of water autonomy to a vast majority of Bulawayo residents, ZANU PF lost its legitimacy, even with many of its former supporters. As Hossain and Kalita (2014) have observed, in a context where the responsibilities assigned to public authorities are informed by formal and electoral accountabilities, elections not only are a means of holding leaders to account, but they also become strategic forms of collective action.

In 2009, the Inclusive Government, a national coalition between MDC and ZANU PF (2009–2013), reversed the policy to centralize the reticulation of water (Mabiza 2013). 'Our water' remained in the hands of the city. Throughout the protests, the local city council had acted as the custodian of water – the precious life-source of the city. It had reminded the national government of its obligation to guarantee local autonomy in managing water so that the city could stay alive.

At the same time, the council had thereby adopted the role of water *provider*, thus shifting its position in the framework of the moral economy. A few years later, it was held accountable on those very grounds when, in concert with the national government, it announced the installation of prepaid water meters across the city as a measure to 'improve revenue collection' (BPRA 2015). A second wave of protests unfolded, this time led by a city-wide voluntary association, the Bulawayo Progressive Residents Association (BPRA). The association had been formed earlier by a group of concerned residents – youth, women, workers and environmental activists – who insisted that even after the initial victory, it was important to remain vigilant and to defend their city's water against all onslaughts.

Claiming water as a human right: The protests of 2013–2015

Just as in the previous attempt to introduce water control by ZIMWA, the official rationale for the introduction of prepaid water meters (PPWM) was a techno-political solution to *improve* supply and access to water. This was to be achieved through a move towards becoming a 'smart city', i.e. establishing efficient city management through technology-based systems (Datta 2016). Yet, the residents of Bulawayo fiercely rejected this intervention, framing their protests using the language of indignation and need. Between 2013 and 2015, a loose coalition of civic actors, named the 'Right to Water Campaign', headed by the BPRA, led the resistance against the installation of prepaid water meters. This time, the protesters directed their concerns primarily at the Bulawayo City Council. The council had shifted from a representative of local concerns against the national government to an emblem of the local state and, ironically, even a representative of the national government.

In this section, we demonstrate how this shift in positionalities took place and how protesters rephrased their claims from an emphasis on local autonomy to matters of dignity, human rights and basic needs. While the first protest had relied largely on elected political representatives, the second protest took to the streets and mobilized many more people. Putting forth the aspect of dignity, protesters demanded recognition – not as a matter of charity, but as a matter of rights. Water, they insisted, was a common good that should not be commoditized or measured by market indicators; rather, it should be controlled by the people. Calling upon the local council to protect water resources as well as the citizens' livelihoods, protesters articulated their political subjectivity in a framework of moral relations of reciprocity. The fact that the city council appeared to ignore those affective links by which it had previously propelled itself enraged the protesters even more and sharpened their resistance.

Led by the BPRA and the Right to Know Campaign, and comprised of sixteen civil society organizations, they opposed the city's initiative on three fronts: (i) the decision was made without consulting residents; (ii) the water meters were an 'affront to the notion of water as a human right'; and (iii) the meters had dire social, economic and environmental implications (BPRA 2015).

Local democracy and governing the commons

One of the arguments was that the Bulawayo City Council had not consulted residents before making the resolution to install prepaid water meters. This decision was not only a breach of trust, but it also violated the citizens' constitutionally protected right to participate in decision making and government commitments to effectively engage communities. Clearly, the introduction of prepaid water meters would have huge implications on citizen-state relations in terms of welfare and social rights. BPRA insisted that such a decision should not have been made haphazardly just because the council was attracted to a vendor who had sold them an idea (BPRA 2015).

In making this argument, BPRA cited the scholarly work of von Schnitzler (2014) to substantiate its reasons against prepaid meters. In her discussion of techno-politics and citizenship in post-apartheid South Africa, von Schnitzler demonstrates how water meters turn the reciprocal relationship between the state and its citizens into a fragile consumer relation that does not satisfy citizens' claims to basic rights, especially when it comes to common goods such as water. BPRA turned this argument towards the local council, thereby challenging the latter's self-image as a social democratic outfit and different from the ruling party, which it had always portrayed as neoliberal and dictatorial.

Prepaid meters and the right to water

A second argument was that access to potable water is a human right. The use of prepaid meters would deny citizens (especially the poor, who were likely to be more affected) the right to live a healthy life. To this end, activists cited international and local statutes such as the 2002 General Comment adopted by the United Nations' Committee on Economic, Social and Cultural Rights (United Nations 2003) and the Zimbabwean Constitution's Section 77 that guarantee the right of access to water. Of particular importance in reference to the country's constitution was the state's obligation to 'take reasonable legislative and other measures' (Constitution of Zimbabwe 2013: 38). The residents argued that the decision to install prepaid meters was unreasonable as it was likely to force the poor to use unsafe water. Elaborating on their rights-based claims, BPRA even invoked the World Health Organization's standards (WHO 2003). WHO states that to achieve hygiene, each person needs 50 litres per day, and to lead a healthy life, each person needs 100 litres per day. BPRA argued that the PPWM decision would leave the city far off the mark of meeting these standards.[14] Protesters also referred to the social, economic and environmental implications of lacking water. In their view, prepaid meters would undermine the social contract between residents and the local authority, which was supposed to be anchored on respect and fulfilment of human rights. Focusing on the technology itself, BPRA argued that residents would be limited to interactions with their meters rather than seeking recourse via a dispute mechanism in instances where meters were faulty and water units ran out.

The carefully prepared BPRA statements marked a major change in the dynamics of protest. While the first protest had enacted locality in relation to historical memory and national politics, the point of reference now shifted to a globally circulating framework of human rights. This did not imply an absence of locality. On the contrary, it was an indicator for new forms of political subjectivity that are not necessarily bound to formal citizenship status, as famously proposed by Hannah Arendt (1955), but rather are articulated around social rights (Marshall 1998; Turner 2009). This implies that multiple bodies of authority may be evoked to substantiate a claim (Krause and Schramm 2011): from the work of an anthropologist to the national constitution, from WHO statistics to global activism around the protection of the commons.

Prepaid water meters and economic, social, and environmental implications

The third objection was directed towards the city's accusation that the residents were not paying water bills because they were not willing to, rather than not being able to. At the time the decision was made, Zimbabwe had undergone economic challenges for over a decade. Residents argued that BCC should have considered this socio-economic context. More poignantly, they asserted that prepaid meters would magnify inequality and poverty as the commoditized water would be accessible only to the wealthy segment of the population:

> It is fine for the rich people to use as much water as they can because they can afford it, while it is right for the poor people to fail to meet their basic water requirements because they cannot afford it. Prepaid water meters force poor families to abandon gardening activities that are essential for their subsistence while rich people have the leeway to water lawns and flowers and abuse [*sic*] water for recreational purposes. They are thus discriminative and penalize the poor.
>
> (BPRA 2015: 17)

The council's techno-political decision was thought to be ill-informed and thus further entrenched poverty. This point is related to the PPWMs' possible negative effects on public health – for instance, the risk of waterborne diseases such as cholera – because once the poor are forced to use less water, they resort to polluted sources. The policy also came against the backdrop of controversy and political failures in the wider region. BPRA (2015) noted negative implications for public health, citing cases of cholera outbreaks such as Madlebe in KwaZulu-Natal in South Africa after the introduction of prepaid water meters in 2000.

The mooted prepaid water meter system was also likely to affect the progress made in terms of gender equality by returning girls and women to being 'alternative service providers', thus depriving them of life chances and opportunities. The most affected were likely to be pregnant women, those with young children and those going through monthly menstrual cycles. On a broader scale, the PPWMs bore the risk to negatively affect communities' social capital and thus atomize and depoliticize residents. BPRA (2015) cited Dawson's findings that in Soweto, Johannesburg, friendships and family networks had been severely disturbed after the introduction of prepaid water meters in 2000.

Residents taking action[15]

Residents' groups employed different forms of collective action such as petitioning, meetings, marches and even threats of violence if the policy went ahead without their consent.[16] From 2013, the time the resolution was made, to 2015, the Right to Water Campaign brought residents together to participate in activities aimed at encouraging the Bulawayo City Council to rescind their decision. Three peaceful demonstrations-cum-marches were held in the city centre, and specifically at the City Hall. These were preceded by massive groundwork that saw activists conducting house meetings, door-to-door campaigns, media articles and press releases as well as letters to the individual council members, senior technocrats and MDC leaders sharing research-based policy positions. All were called upon to rally behind the protests, which turned out to be one of the biggest protests in the city's post-colonial history.

Activists also kept a close eye on the council: all minutes of council meetings were carefully scrutinized, and members of BPRA in particular personally observed council meetings. Pressure was applied and maintained on them, both individually and collectively. Platforms for engagement between councillors and the residents were also put in place by different civil society organizations, where residents directly conveyed their reservations against the PPWMs.

On placards and in songs and slogans, protesters invoked the immorality of the decision and referred to bad decisions of the past that had negatively affected the people of Bulawayo. People carried placards like 'Prepaid water meters are like Gukurahundi' – once again evoking the emotional weight of the collective memory of a murderous assault on the local community. For the MDC-led Bulawayo City Council to be equated to an uncaring ZANU PF government was a huge affront. The council, which had once accused the central government of wanting to destroy the city, was now itself under attack for deliberately destroying it by taking a decision that put the lives of residents and the city's socio-political cohesion at risk.

During the final demonstration in August 2015, the Right to Water Campaign showed force by bringing thousands of residents onto the streets once more, marching from the City Hall through the city to end with a rally at the Stanely Square in Makokoba. The venue was chosen not only for its proximity to the city, but also for its significance in the nationalist mobilizations of the past by PF-ZAPU led by the city's legendary figures such as Joshua Nkomo. To counteract the managerial and technocratic logics of the administration, the organizers ensured the participation of all residents in the protests, especially the most vulnerable such as old women, unemployed youth and street vendors. *Their* right to water was a moral obligation, not just an element in a technical game.

The protesters stayed clear of the legal route even though they employed the registers of human rights. Rodrick Fayayo, the former BPRA Coordinator, explained that the Right to Water Campaign did not approach the courts:

> because we had seen the risk of failure in the courts, given their lack of independence. In fact, other organizations, for instance the Zimbabwe Lawyers for Human Rights, had failed in its attempts to stop the erection of tollgates in the cities through the courts. Again, examples from the neighbouring South Africa were telling, the Anti Privatisation Forum's case, for example, had been dismissed by the ConCourt in that country. So we could not repeat the same mistakes. Our power is in mobilization, the numbers, the anger of the residents and their ability to punish the elected leaders through elections.
> (Interview, 13 March 2019)

In this statement, the force of protest as a political act clearly comes to the fore – through affective mobilization, residents gained a voice and became visible as political subjects to be reckoned with: as a massive force on the streets, and as the sovereign during election time. If the authorities would not fulfil their moral and social obligation to the people, they would be punished.

At the beginning of 2016, the council finally made a veiled withdrawal of the policy, climbing down from compulsory installation of prepaid water meters to resolving to install the PPWMs on a voluntary basis (Sunday News 2016). Once again, the citizens of Bulawayo had succeeded in fighting off an unwanted policy relating to water. 'Our water' remained in the hands of the municipality, which was expected to care for its citizens and provide better services. Outsourcing the responsibility to the residents themselves was not an option.

Residents made it clear that neither the legitimacy of the national state nor that of the local council was a given, but that it was attached to the respective authority's capability to provide adequate services. Moreover, they also insisted that the city (and not the nation) was their frame of claim-making – thereby appealing to very specific, even personal, relations and responsibilities. Finally, protesters were not content with the role of passive recipients but linked their claims to broader questions of democratic political participation.

Conclusion

This chapter has addressed post-independence struggles over social values, citizenship rights and political legitimacy through close attention to the politics of locality and shifting dynamics of moral economies. In our discussion of water protests in the city of Bulawayo, we have underlined how important it is to pay attention to the specific articulation of protest relations. Positionalities in the moral economy of protest are not fixed. Even if the object of contestation in both cases was water, the precise scope of the common good was contingent on wider historical and political relations. Whereas the emphasis in the first protest was on locality and ownership, i.e. *our* water, the second protest focused on our *water*, i.e. the resource itself as a means of subsistence.

Locality was not fixed either, as it entailed deep layers of memory and various points of reference – from the historical significance of the pre-colonial Bulawayo to the city's important role in the liberation struggle; from the violence of post-colonial power struggles to the resonances of creative resilience. Likewise, the claim to the common good was fluid – it expanded and contracted according to the specific constellation in which it was articulated. In the first instance, the affirmation of local identity and the city's autonomy determined the contours of the common good. The state was called upon as a responsible provider but also as an antagonist in the struggle for political legitimacy. Consequently, the asset at stake was not only the water itself, but also the government's financial support and the freedom of the city to manage its own affairs. In other words, the major resource and point of contestation was political autonomy within the realm of citizenship.

In the second incident, the addressee of the protests was the local council which had failed in its expected role as a protector of the city's interests and as a service provider. The residents who took to the streets felt betrayed and regarded the commoditization of water as another form of sell-out, this time to a neoliberal market logic. In their eyes, the local council (just like the national government before), had lost its legitimacy to represent the people. The protest was therefore an alternative assertion of residents' political subjectivities. In this affirmation of their local political voice, activists drew on the globally circulating discourse around human rights, dignity and the commons. Theirs was not an isolated protest, but a response to larger political and economic transformations. In caring about water and showing solidarity across party and other divisions, the protesters enacted a moral economy that was deeply entrenched in a sense of community, place and entitlement, while also holding up a promise to another future.

Notes

1 Bulawayo was the industrial hub for Southern and Northern Rhodesia, especially in terms of railway transport. Leaders such as Joshua Nkomo, who led the ANC, NDP and subsequently ZAPU, were labour unionists in Bulawayo.

2 For more extensive discussions about the analytical value of the concept, see Edelman (2005), Fassin (2009), Friberg and Götz (2015) and Carrier (2018).

3 Also important to note is that the Rhodesian Water Act of 1976 introduced private ownership of water and attached water and land rights. This meant that one had to hold a title to land in order to acquire water rights. Black people in the reserves who had been deprived of land titles could thus not claim unfettered access to water. The Act also prioritized those who already held water rights, thus disadvantaging new applicants (Jaspers 2001, in Youde 2010).

4 *Gukurahundi* denotes the state-orchestrated massacre of at least 20 000 Ndebele-speaking civilians in the Matabeleland and Midlands provinces between 1983 and 1987. The Mugabe-led ZANU PF regime accused PF-ZAPU of plotting against the government. The government then deployed a North Korean-trained Fifth Brigade militia, which consisted of Shona-speaking ex-combatants of the military wing of ZANU PF, to the Matabeleland and Midlands provinces to subdue the alleged 'dissidents'. The Fifth Brigade committed killings, mass detentions, disappearances, torture and rape against PF-ZAPU officials, members of that party's former military combatants and Ndebele-speaking civilians who were accused of aiding 'dissidents'. The Brigade also claimed to be avenging the pre-colonial Ndebele raids on

Shona communities. The overt violence ended in 1987 after a truce between ZANU PF and PF-ZAPU. However, the perpetrators have not been held to account. For more details, see Catholic Commission for Justice, Peace in Zimbabwe and Legal Resources Foundation (Zimbabwe) (1997).

5 Between 1980 and 1987, Robert Mugabe was Prime Minister while Canaan Banana was the ceremonial President. Constitutional changes in 1987 created an executive Presidency, replacing the ceremonial post of President and incorporating the post of Prime Minister. Mugabe became Zimbabwe's first executive President at the end of 1987.

6 In the 1980s, ZANU PF regarded the electoral trend in Bulawayo as an injunction on its plans to establish a one-party state. Later, it perceived the ongoing support for the opposition as a threat to its hold on state power (Cheeseman and Tendi 2010).

7 In terms of the revenue, the city was said to be deriving 40% of its revenue from water and using it to subsidize other social services.

8 Payne had been active since the 1980s and 1990s. In 1992, he undertook a lone campaign to highlight the water problems in Bulawayo. He pushed a wheelbarrow carrying 210 litres of water from the Zambezi River to Bulawayo and then to Gwanda (more than 500 kilometres), emptied the drum and then continued his journey with an empty wheelbarrow to the national parliament in Harare (another 500 kilometers). His activism, clothing and vehicle inscribed with the words 'Zambezi Water for Matabeleland', were a constant reminder of the city's water problems. He died in 2014.

9 Harare and other areas under ZINWA had not seen improvements in water management. In some instances, things had gotten worse. See Youde (2010) and Kamete (2006) for a detailed discussion.

10 It should be noted that the takeover of municipal water supplies by the national government through ZINWA, coupled with economic mismanagement at the time, are seen to have 'worked in tandem to allow the 2008–2009 cholera outbreak to flourish' (Youde 2010). This epidemic affected almost the entire country, with the exception of Bulawayo, where infection rates remained low.

11 Interview by Mmeli Dube with Mbuso Fuzwayo, leader of Ibhetshu LikaZulu, a civil society organization based in Bulawayo. The interview was conducted on 26 February 2019. Ibhetshu LikaZulu seeks justice for Gukurahundi victims, as well as truth telling and commemoration/memorialization of the massacre. It also stands for peace and prevention of further persecution and marginalization of the people of Matabeleland by the ruling elite. Fuzwayo was directly involved in the campaign against ZINWA taking over the city's water management duties and infrastructure.

12 Ibid. See also Masuku and Mudzviti (2007) for a thorough discussion of the protest actions (demonstrations, disruption of meetings, press statements, etc.).

13 Re-urbanization is a term used by Kamete (2006) to refer to ZANU PF's attempts to win votes in the urban areas after the emergence of the Movement for Democratic Change, a political party that won virtually all parliamentary and local government elections in major cities and towns.

14 In von Schnitzler's (2014) case study, the South African state cited these very standards in order to justify its water policy – and activists opposed this form of quantification as it did not mirror the actual needs of people. This shows again that positionalities in the moral economy of water are not fixed but need to be situated in concrete relations.

15 The material in this section is based on the BPRA (2015) policy position, interviews with former BPRA Coordinator Rodrick Fayayo, as well as personal recollections by Mmeli Dube, who led the Right to Water Campaign in 2015.

16 In particular, Mbuso Fuzwayo, the leader of Ibhetshu LikaZulu, which was part of the coalition, told a meeting held on 17 August 2015 that, 'if the Council goes ahead with installation, against the residents, we will smash those meters!'

References

Agere, S.T. 1987. 'Progress and Problems in the Health Care Delivery System', in I. Mandaza (ed.), *Zimbabwe: The Political Economy of Transition, 1980–1986*. Dakar: CODESRIA, pp. 354–377.

Arendt, H. 1955. *Elemente und Ursprünge totaler Herrschaft* [The Origins of Totalitarianism]. Frankfurt: Europäische Verlagsanstalt.

Baker, J. 2012. 'Water Supply Challenges in Bulawayo, Zimbabwe', *GIS in Water Resources*. Retrieved 17 May 2019. www.ce.utexas.edu/prof/maidment/giswr2012/TermPaper/Baker.pdf.

Bolding, A., E. Manzungu and P. van der Zaag. 2000. *A Realistic Approach to Water Reform in Zimbabwe*. Harare: University of Zimbabwe (UZ) Publications.

Bratton, M. 2009. 'Authoritarian Resilience and State Failure: Lessons from Zimbabwe'. Paper presented at the *2009 Annual Meeting of the American Political Science Association*, Toronto, Ontario, Canada.

Bulawayo Progressive Residents Association (BPRA). 2015. *Controversies Surrounding Prepaid Water Meters: The Case of Bulawayo: A Survey Contextualized Within the Framework of the National Water Policy in Zimbabwe*. Bulawayo: Directory Publishers Ltd.

Carrier, J.G. 2018. 'Moral Economy: What's in a Name', *Anthropological Theory* 18(1): 18–35.

Catholic Commission for Justice, Peace in Zimbabwe and Legal Resources Foundation (Zimbabwe). 1997. *Breaking the Silence, Building True Peace: A Report on the Disturbances in Matabeleland and the Midlands, 1980–1988*. Harare: Catholic Commission for Justice and Peace in Zimbabwe.

Cheeseman, N., and B.M. Tendi. 2010. 'Power-Sharing in Comparative Perspective: The Dynamics of "Unity Government" in Kenya and Zimbabwe', *Journal of Modern African Studies* 48(2): 203–229.

Constitution of Zimbabwe. 2013. Last accessed 21 July 2020. http://www.veritaszim.net/sites/veritas_d/files/Constitution%20of%20Zimbabwe%20Amendment%20%28No.%2020%29.pdf.

Datta, A. 2016. 'Three Big Challenges for Smart Cities and How to Solve Them'. *The Conversation*. Last accessed 14 February 2019. https://theconversation.com/three-big-challenges-for-smart-cities-and-how-to-solve-them-59191.

Edelman, M. 2005. 'Bringing the Moral Economy Back In … to the Study of 21st-Century Transnational Peasant Movements', *American Anthropologist* 107(3): 331–345.

Fassin, D. 2009. 'Moral Economies Revisited', *Annales: Histoire, Sciences Sociales* 64(6): 1237–1266. Last accessed 22 May 2019. https://www.cairn.info/revue-annales-2009-6-page-1237.htm.

Friberg, K., and N. Götz. 2015. 'Introduction to the Thematic Issue "Moral Economy: New Perspectives"', *Journal of Global Ethics* 11(2): 143–146.

Gwebu, T. 2002. 'Urban Water Scarcity Management: Civic vs. State Response in Bulawayo', *Habitat International* 26: 417–431.

Hossain, N., and D. Kalita. 2014. 'Moral Economy in a Global Era: The Politics of Provisions During Contemporary Food Price Spikes', *Journal of Peasant Studies* 41(5): 815–831.

Jaspers, F.G. 2001. 'The New Water Legislation of Zimbabwe and South Africa – Comparison of Legal and Institutional Reform'. *International Environmental Agreements* 1(3): 305–325.

Kamete, A.Y. 2006. 'The Return of the Jettisoned: ZANU-PF's Crack at 'Re-Urbanising' in Harare', *Journal of Southern African Studies* 32(2): 255–271.

Krause, K., and K. Schramm. 2011. 'Thinking through Political Subjectivity', *African Diaspora* 4(2): 115–134.

Mabiza, C. 2013. *Integrated Water Resources Management, Institutions and Livelihoods under Stress*. London: CRC Press.

Majaka, N. 2013. 'Mugabe Disowns Harare, Bulawayo'. *Daily News Live*. Last accessed 14 April 2019. https://www.dailynews.co.zw/articles/2013/08/26/mugabe-disowns-harare-bulawayo.

Mamdani, M. 1996. *Citizen and Subject: Contemporary Africa and the Politics of Late Colonialism*. Princeton: Princeton University Press.

Marshall, T.H. 1998. 'Citizenship and Social Class', in G. Shafir (ed.), *The Citizenship Debates*. Minneapolis: University of Minnesota Press, pp. 93–112.

Masuku, F., and J. Mudzviti. 2007. 'Bulawayo Residents Walk Out on ZINWA-Related Budget Meeting'. *Zimbabwe Coalition on Debt and Development (ZIMCODD)*. Last accessed 28 June 2018. http://archive.kubatana.net/html/archive/locgov/071012zim-codd.asp?sector=LOCGOV&year=2007&range_start=1.

Mate, R. 2018. 'Social Policy and Social Spending in Zimbabwe: 1980–2015', *UNRISD Working Paper* (8). Geneva: UNRISD.

Mpofu, B. 2010. 'No Place for 'Undesirables': The Urban Poor's Struggle for Survival in Bulawayo, Zimbabwe, 1960–2005', PhD Dissertation. Edinburgh: University of Edinburgh.

Musemwa, M. 2006. 'Disciplining a "Dissident" City: Hydropolitics in the City of Bulawayo, Matabeleland, Zimbabwe, 1980–1994', *Journal of Southern African Studies* 32(2): 239–254.

Musemwa, M. 2008. 'The Politics of Water in Post-Colonial Zimbabwe, 1980–2007', *African Studies Centre*. Netherlands: University of Leiden.

Musemwa, M. 2010. 'From "Sunshine City" to a Landscape of Disaster: The Politics of Water, Sanitation and Disease in Harare, Zimbabwe, 1980–2009', *Journal of Developing Societies* 26(2): 165–206.

Nyathi, K. 2008. 'ZINWA Shelves Bulawayo Takeover'. *Zimbabwe Standard*. Last accessed 21 May 2019. https://www.ircwash.org/news/zimbabwe-zinwa-shelves-bulawayo-takeover.

Nyoni, M. 2007. 'Bulawayo Faces Water Crisis'. *Mail & Guardian*. Last accessed 3 July 2018. https://mg.co.za/article/2007-09-17-bulawayo-faces-water-crisis.

Ranger, T.O. 2007. 'City versus State in Zimbabwe: Colonial Antecedents of the Current Crisis', *Journal of Eastern African Studies* 1(2): 161–192.

Resnick, D. 2014. 'Urban Governance and Service Delivery in African Cities: The Role of Politics and Policies', *Development Policy Review* 32(1): 3–17.

Rukuni, C. 2007. 'Take-Over of Water Supplies in Bulawayo a Travesty – MDC'. *Financial Gazette*. Last accessed 22 July 2020. https://allafrica.com/stories/200710251239.html.

Rusinga, F., and A. Taigbenu. 2005. 'Groundwater Resource Evaluation of Urban Bulawayo Aquifer', *Water SA* 31(1): 23–34.

Scott, J. 1976. *The Moral Economy of the Peasant*. New Haven: Yale University Press.

Sibanda, G. 2003. 'Zinwa, Stay Away from Byo', *Zimbabwe Independent*, 6 June 2003.

Siméant, J. 2015. 'Three Bodies of Moral Economy: The Diffusion of a Concept', *Journal of Global Ethics* 11(2): 163–175.

Sunday News. 2016. 'No More Pre-Paid Water Meters', *Sunday News*. Last accessed 21 July 2020. https://www.sundaynews.co.zw/no-more-pre-paid-water-meters/.

Thompson, E.P. 1971. 'The Moral Economy of the English Crowd in the Eighteenth Century', *Past and Present* 50: 76–136.

Turner, B.S. 2009. 'T.H. Marshall, Social Rights and English National Identity', *Citizenship Studies* 13(1): 65–73.

United Nations. 2003. 'General Comment No. 15: The Right to Water (Arts. 11 and 12 of the Covenant) Adopted at the Twenty-Ninth Session of the Committee on Economic, Social and Cultural Rights, on 20 January 2003 (Contained in Document E/C.12/2002/11). Office of the High Commissioner for Human Rights'. Last accessed 21 July 2020. https://www.refworld.org/pdfid/4538838d11.pdf.

Von Schnitzler, A. 2014. 'Performing Dignity: Human Rights, Citizenship, and the Techno-Politics of Law in South Africa', *American Ethnologist* 41(2): 336–350.

World Health Organization (WHO). 2003. *The World Health Report 2003 – Shaping the Future*. Geneva: WHO.

Youde, J. 2010. 'Don't Drink the Water: Politics and Cholera in Zimbabwe', *International Journal* 65(3): 687–704.

7 The social contract, the state and Adivasi protests against large-scale mining in India

Ranjita Mohanty

Introduction

Recent history has seen an intense resistance to large-scale mining by the Adivasi,[1] whose subsistence resource of land has been encroached upon by mining companies for economic growth and profit. At the core of the conflict between the poor and the mining companies lies the issue of ownership, rights and control over land, including forest land, that provide livelihoods and habitat to the Adivasi. This chapter discusses two cases of Adivasi protests to large-scale mining by multinational mining companies in Odisha – the protests against Utkal Alumina International Limited (UAIL), a conglomerate of mining companies from India, Norway and Canada; and the protests against Vedanta Alumina, an Indian multinational company headquartered in London. The chapter situates Adivasi protests in the context of the promises of a social contract between the state and citizens in the wake of freedom from colonial rule in 1947. The social contract makes the state responsible for the socio-economically vulnerable populace. Economic development and social justice are the hallmarks of the contract. The promises of the social contract put in place through state policies, legislations, institutions, and a written constitution make the state the provider, protector and guarantor for rights of its poor populace. The social contract provides strength and legitimacy to Adivasi protests to secure their land rights when the state violates the contract by siding with mining companies. The breach of the promises of the social contract from the side of the state makes the Adivasi resistance necessary to pressure the state into honouring the social contact.

On 16 December 2000, three Adivasi died in a remote village called Maikanch in the state of Odisha – located in the eastern part of India – in a police shooting that took place to intimidate people resisting the UAIL that had proposed a plan to acquire their land for its mining project. Thousands of Adivasi in the area organized a protest march on 31 January 2001 and vowed that they would not leave their land, even if it meant being killed. Although industrial projects had displaced the Adivasi from their land since the planned development[2] began in the early 1950s, the conflict had not escalated to the extent that people had to die. A striking feature of economic growth, particularly in the mining sector since India liberalized and adopted a neoliberal market economy in the early 1990s, is the manner in which it is pursued. These deaths[3] were not only the defining moments of the conflict between the Adivasi subsistence economy based on natural resources and the market economy that wants to appropriate these resources for profit; they were also indicative of the role that the state plays in subverting Adivasi land rights. Fraught with different meanings and practices associated with the use and ownership of land as well as rights over land, the conflicts have reached a point where the people and their land that they die for have become inseparable.

The Adivasi population makes up 8.6% (104 million people) of the total population of India; close to 10% of this population live in the state of Odisha and constitute 22% of Odisha's total population (Government of India 2011). The Adivasi communities mostly practice agriculture, though some are also herders and hunter-gatherers. The Adivasi are highly dependent on forests for food – including fruits, nuts and roots – and wood for fuel and house construction. The Constitution of India categorizes them as Scheduled Tribes, minority groups that are socially and economically distinct and vulnerable. The Constitution also makes special provision for them through affirmative action. Adivasi areas are given special status in the Constitution as Scheduled Areas having special administrative arrangements in states with a large Adivasi population. The Panchayat[4] Extension to Scheduled Area Act 1996 has vested power in local governance institutions of panchayat and village assembly (called *gram sabha*) to make decisions on social and economic development and preserve the culture and tradition of the Adivasi. Furthermore, there are special grants for education and welfare provided to the Adivasi.

India had followed a mixed economy in which the state-controlled public sector played a larger and significant role until the early 1990s, when the country embraced a new economic order premised on liberalization, privatization and globalization, popularly known as LPG. India liberalized its economy and adopted a free market economy by substantially expanding its private sector. The country also opened up its markets and facilitated the entry of multinational companies to increase investment in the domestic economy. The Tenth Five Year Plan (2002–2007) outlined the potential of the private sector in propelling economic growth. The plan carved out a much bigger role for private industries and minimized the role of state intervention to mere facilitation (Planning Commission 2001). International mining had emerged as a lucrative project to drive economic growth, and India wanted to exploit the rich mineral deposits in the country.

The regions inhabited by the Adivasi are rich in mineral deposits such as coal, bauxite and iron ore. These deposits have led to both public and private mining companies making inroads into Adivasi land. Large-scale mining operations have displaced the Adivasi from their homestead and agricultural land as well as from the forest land that has traditionally been inhabited by forest-dwelling communities. Therefore, they have lost both their individual land and their communal land.

This chapter argues that the Adivasi claims on land rights in contemporary times are based on the promises of a social contract between the state and its citizens, which was mutually constructed when India became independent from colonial rule. The contract became a basis for expectation from the state; it set rules and regulations as well as norms and standards for the state to follow. The social contract has suffered fractures both big and small, but the promises of the contract have not been completely abandoned by the state, nor have they stopped shaping people's expectation from the state. The conflict between the Adivasi subsistence economy and profit-oriented mining companies over ownership and rights to land resources points to the role of the state in diverting Adivasi land to the mining corporations and thus constitutes a breach of the social contract the state has with the Adivasi. The protests show that people believe that they can make the state respond to their claims for land rights.

Adivasi and their land: A history of exclusion, exploitation and impoverishment

In the Adivasi *lebenswelt* (lifeworld), the meaning of land has various material, cultural and epistemological connotations. A piece of land is not only a source of livelihood; it is also the Adivasi's ancestral land, a symbol of family and community ties and a place where their deities reside and continually bestow blessings for wellbeing. Land, for the Adivasi, is not always private property – land is also a communal and social good. Land has a special significance in the Adivasi worldview that revolves around nature, and the knowledge about nature is the wisdom that sustains their lives. Their identity is closely linked to the place where they have lived for generations. The loss of land has terrifying consequence for the Adivasi. Women who protested against UAIL said:

> For us the land, forests, hills and the rivers of Kashipur are the source of livelihood and also our Gods. We worship these lands … If you take away our natural resources from us, then we cannot live and the money given to us for the land is of no use. We only know about our land. What will we do with money?
>
> (Patibandla 2009: 12)

Bula Miniaka, an Adivasi woman said:

> Our life, culture, and employment; everything revolves around the soil … In the month of *Ashadh* (signifying the rainy season), we have *Kulimaru parab* (Adivasi festival), we get the root of the crop home and worship it. In the month of *Bhoda* (the following month) we worship maize. … We will get jobs, but who will do the cultivation? People will eat money or what?
>
> (Ibid.: 4)

Sikaka Kunji, an elderly woman who protested the Vedanta mining project, pointed to the Adivasi knowledge system when she said:

> We are not educated but we have the knowledge needed to live wholesome lives and we are not fools.
>
> (Jena 2013: 15)

The Adivasi have been rendered poor and resourceless, first by the princely rulers and the colonial administration, and later by the Indian state and its power structures post-independence.

Taxation during the colonial period

The farmers were taxed both by the princely states as well as by land governance systems introduced by the colonial administration. The burden of rent drove the Adivasi into a never-ending circle of debt from moneylenders (called *sahukars*), most of them outsiders who came to Adivasi areas when the colonial administration expanded the communication network. While indebtedness impoverished the Adivasi and drove them to work as wage labourers in the tea gardens of Assam (Mohanty 1997), the loss of land pushed them deeper into forests and mountains (Pati 2013). It also gave rise to *Gothi*, a practice of bonded labour that continued until 1976, when the Bonded Labour Abolition Act was passed (Mohanty 1997). The legislation could not abolish slavery completely, and it continued in hidden forms due to the fact that the bonded labourers did not identify themselves because they feared the consequences of exposing their masters (Ibid.).

Transfer of Adivasi land to non-Adivasi

The land management systems of the colonial administration were abolished after independence. Nonetheless, the threat of dispossession of land and impoverishment remained as outsiders and wealthy upper castes from other parts of Odisha came to Adivasi motivated by the prospects of grabbing land, moneylending and working as contractors. To control the transfer of Adivasi land to non-Adivasi, the Orissa Scheduled Area Transfer of Immovable Property Act was passed in 1956. The legislation allowed the transfer of land legally owned by the Adivasi to non-Adivasi with the permission of a designated government authority. Taking advantage of the illiteracy of the people who were not conversant with legal and administrative procedures, the land transfer continued through the nexus of moneylender, contractor and bureaucrat, most of them non-Adivasi outsiders. This legislation was amended only in 2002. The new legislation barred the transfer of land except as gifts or as mortgage to financial institutions for the purpose of agriculture or self-employment. Furthermore, the amended legislation allowed the exchange of land for public purpose (Ambagudia 2010).

As part of the state's social justice agenda, land reforms were also initiated to provide land free of cost to the poor. Odisha enacted this legislation in the mid-1970s. The Land Ceiling Act prescribed a ceiling for individual/family holdings, and the surplus land thus obtained was to be distributed among the marginal farmers and landless labourers. Although land was distributed to the Adivasi, in some reported cases, the bonded labourers were forced by their masters to permit them to cultivate their land (Mohanty 1997).

Restricted access to forests

The state, together with its modern institutions of law and private property, began regulating Adivasi access to forest land during the colonial period. The colonial state regulated access to forests by marking large patches as 'reserved forests', where entry was not permitted, and by creating conditions for access to 'protected forests'.

After independence, the state continued using the colonial systems of control over forests (Guha and Gadgil 1995; Pathak 1994).

Reserved forests remained the exclusive domain of forest departments, which barred the Adivasi from even collecting firewood. Protected forests provided limited entry, fines were levied for breaking the rules and bribes were taken by junior forest officials to allow what was not permissible by law (Mohanty 1997). The Adivasi customary rights on forests were not recognized, as there were no legal documents to prove their ownership. Even though the government of Odisha introduced a number of forest regulations after independence, the regulations have been used as instruments to control forests and exclude the Adivasi rather than restore their customary rights. Although joint participatory forest management was adopted, the forest bureaucracy still exercised its authority (Satpathy 2017). It was only in 2006 that the Forest Rights Act was passed as a historical step towards giving the Adivasi their rightful share in the forests that have been their traditional home and a major source of their livelihood.

Displacement by industrial projects

A large number of Adivasi have been displaced from their land and habitat by development projects, particularly by big dams and mining (Baboo 1991; Fernandes 1991; Mohanty 1997; Sahoo 2015; Stanley 1996). Dams such as Machkund in 1954, Upper Kolab in the mid-1970s and Indravati in 1978 have augmented irrigation and hydroelectricity, but the construction of these dams has also displaced people. The second type of development projects that have displaced the Adivasi are large-scale mining projects. It began with the National Alumina Company Limited – a public-sector company – in the early 1980s. Once India liberalized its economy, the private sector was given access to Adivasi land and forests for mining operations. Besides displacement resulting from direct mining operations, mining infrastructure and campuses[5] owned by the companies have also displaced people.

The resettlement and rehabilitation of displaced people have remained slow, *ad hoc* and inadequate (Mishra 2002; Mohanty 2005; Ray and Saini 2011). There are also many caveats and shortcomings in the form that resettlement and rehabilitation have taken. There is usually no prior information or discussion with people regarding the acquisition of their land. Cash compensation is never adequate to compensate for what people lose in terms of their livelihood and habitat. The idea of land for land, as good as it may sound, has presented immense challenges in terms of the availability of good-quality land and relocating people as close as possible to where they have lived all their lives. Besides this, those who do not have legal rights to their land, such as tenants, herders and landless agricultural labourers, get no land compensation except for homestead land. Due to the levels of illiteracy among the Adivasi and the manner in which displacement takes place, people have been unable to negotiate a market price for their land.

Displacement has pushed many Adivasi into economic destitution and cultural alienation. The history of economic growth in Adivasi areas has been the history of exclusion, displacement, treachery and violence for the Adivasi. They have lost their land and have become unskilled wage labourers. This partly explains why the contemporary movements have discarded the negotiation for compensation altogether and have refused to sell their land.

Two cases of Adivasi protests against large-scale mining

Resource conflicts between the subsistence economy of the poor and an economy geared towards profit began soon after independence but were not perceived as conflicts at the time. Profit was seen as the creation of national wealth and hardships were thought to be temporary and a contribution to nation building. A new narrative began with the protests that emerged two decades after independence (Bavisker 1995; Bhatt 1991; Guha 1989; Mohanty 2003). This new narrative equated ecology with economy and articulated the concerns of the poor. Poverty began to be understood as the commercial exploitation of ecological resources that had created an enclave of prosperity for a section of population at the expense of the vast majority who survive on the ecological resources of land, water and forest (Agarwal 1985; Bandopadhyay and Shiva 1988; Gadgil and Guha 1994). The local resistance movements have since raised questions regarding the rights over resources, economic growth, marginalization and sustainability of resources that serve as the primary source of subsistence for the poor. In the early phase, the protests had demanded the resettlement and rehabilitation of displaced people. The movements resisting the ruthless neoliberal growth agenda in recent years have been asserting poor people's rights over their land and their livelihood resources.

The land rights of the Adivasi on private as well as common property are currently governed by two major pieces of legislation: the Panchayat Extension to Schedule Area Act (PESA) 1996 (Ministry of Tribal Affairs 1996) and the Scheduled Tribes and Other Traditional Forest Dwellers (Recognition of Forest Rights) Act 2006, popularly known as Forest Rights Act (FRA) 2006 (Ministry of Law and Justice 2007). The PESA vests power in *gram sabha* (village assembly) and prescribes that *gram sabha* will have to be consulted before the acquisition or transfer of Adivasi land for any purpose including development projects and industrial enterprise, as well as before plans are made to resettle and rehabilitate the displaced communities whose land has been thus acquired. The PESA thus makes special provision for the Adivasi under decentralized decision-making and vests them with the right to self-determination. The FRA recognizes the rights of the Adivasi (and other traditional forest dwellers whose rights had not been recorded but who had been residing in forests for generations) over forest land both as individual property rights and collective right over community forests. The FRA not only recognizes the land rights of the Adivasi, it also permits them to use forest products and gives them a stake in conservation of forests including maintenance of biodiversity. The FRA is a corrective measure against the restrictions placed on the forest and land rights, first by the colonial administration and subsequently by the Indian state. This legislation, however, has not prevented the state from diverting both private lands of the Adivasi as well as government-owned forests to mining companies.

The Land Acquisition Act 1894, a piece of colonial legislation that vested immense power in the hands of the state, was used unsparingly to acquire both private as well as public land for 'public good'. The Land Acquisition Act was revised in 2013 and renamed as the Right to Fair Compensation and Transparency in Land Acquisition Rehabilitation and Resettlement Act 2013 (Ministry of Law and Justice 2013). The Act provides for the resettlement for all affected people. It does not, however, provide compensation for the land acquired unless the displaced family/individual has a legal title to the land. This has critical implications for the Adivasi as the regulation of legal title to land is still not complete, even though the FRA has sought to legalize these regulations. The revised Land Acquisition Act 2013 sidesteps the prohibition on the Adivasi land being transferred to non-Adivasi by allowing the acquisition of their land by private companies or for public–private partnership projects. The 2014 Ordinance which amends the 2013 Act merely replaces the words 'private company' with 'private entity'.

Protests against UAIL

The Adivasi had protested against a public-sector mining company – Bharat Aluminium Company – in the late 1980s and were successful in driving the company out of the Gnadhamardan Hills in western Odisha. They had not yet known the ruthlessness of a neoliberal state. The situation changed later when a major Adivasi protest began in the mid-1990s against UAIL. Initially, the conglomerate consisted of two private companies – Norsk Hydro Alumina from Norway and Tata conglomerate from India – as well as the Indian Alumina Company (INDAL), an Indian public-sector company. Later, the Tata conglomerate withdrew and Alcan from Canada joined. During this period, HINDALCO, a private Indian company, purchased INDAL and became part of UAIL. The UAIL entered into an agreement with the government of Odisha in 1993 to mine bauxite in the Kashipur block[6] of Rayagada district in Odisha, a district heavily populated by the Adivasi. The mining project proposed an alumina refinery plant, a captive power plant and active mining, with an estimated project cost of $1 billion (Goodland 2007; Patibandla 2009). By 1995, UAIL had acquired over 2800 hectares of land – with 2153 hectares privately owned land and 712 hectares comprising of government land (Patibandla 2009).

The protests against UAIL were first led by an NGO called Agragamee, and then later by an informal community organization called Prakrutika Sampada Surakha Parishad (PSSP). The movement alleged that the scale of displacement was underestimated by the mining project and that the project would directly or indirectly affect a large number of people and villages than originally estimated by UAIL. The people of Kashipur only came to know about UAIL only when the survey for land acquisition began. By then, the company had already signed an agreement with the government. The Environmental and Social Impact Assessment, a mandatory procedure to get approval from the government, conducted in 1995, was not made public (Goodland 2007). Initially, the movement sent a written petition to the government to withdraw the project. In the later stage, the movement assumed a militant tone and demolished the resettlement colony that was constructed by the company to accommodate the displaced people. The protesters barricaded the road prohibiting the company personnel from entering the area. The PSSP also organized *gram sabhas* of affected villages, where people rejected the UAIL and demanded its withdrawal.

The discourse of the PSSP centred on people's rights over land as specified in the PESA and as a special provision of the Fifth Schedule[7] in the Constitution extended to Adivasi regions. The organization put emphasis on decentralization of decision-making processes and the need to validate the *gram sabha* as legitimate representation of the voice of the people of Kashipur. In its narrative it also rejected the state-led projects as a solution to poverty. As Bhagban Majhi, the convenor of PSSP, articulated:

> We have seen big projects in our district like Indravati project and NALCO project. We have heard about Hirakud and Upper Kolab hydro projects. We have seen deprivation and sorrow of local people caused by these development projects. In our view development means that all should live peacefully and happily. But that some should die for development – we do not require this kind of development.
>
> (Patibandla 2009: 9)

The mining project was carried out by deploying police and as the special squad of Central Police Reserve Force. Even though the Maikanch massacre by the police in 2000 was enquired by an independent commission that confirmed police brutality, the Chief Minister of Odisha, instead of taking action against the police, warned the protesters that forces opposed to the industry would be firmly dealt with (Sarangi et al. 2005: 1313). When the protesters opposed the heavy deployment of police and the plan to set up a police checkpoint and barrack in a village in the area where the resistance was taking place, the police came to the place where hundreds of protesters had gathered and announced, 'Womenfolk, clear the road, otherwise we will rape you' (Ibid.: 1314).

The resistance to the UAIL forced the companies from Norway and Canada to withdraw; even the Tata industries withdrew, leaving only the Indian company HINDALCO to continue the mining. Despite the firm local resistance, by 2005, HINDALCO was able to acquire most of the land it required. Subsequently, it constructed the infrastructure for the project and began mining operations. The extent to which the police and brute force was used to supress the protests is illustrative of the ruthlessness of mining in a neoliberal economy.

In the words of two protesters:

> A people struggling just to survive, fight against injustice with extremely limited means. Governments on the other hand, hand-glove as they are with the market forces, can mobilize immense forces to put down such movements. Corporations and governments know that sooner or later the local community will give in and the government which is supposed to be a democracy of, by and for the people, will claim a great 'victory' having defeated its most marginalised people in their fight for human rights and justice. In the bargain

> it will have 'taught a lesson' to other deprived and marginalised sections who might have similar aspirations!
>
> (Das and Das 2006: 35–6)

The later phase of the movement began with pressing the government and the company to give them fair compensation for their land and jobs in the company. The discontents and demands of the protesters were voiced in two public hearing meetings held on 7 October 2006; one was organized by the government and the other by PSSP. The fact that the protests have continued since then conveys the message that the demands have still not been met after almost a decade and a half. Bhagban Majhi summed up the situation when he said: 'The people were promised jobs and development of the region. Despite resistance, the project was cleared. Now people have been left in the lurch, without livelihood' (Pal 2020).

Protests against vedanta mining

The site for another mass protest is yet another Adivasi-populated district called Kalahandi in Odisha, where people have been opposing the bauxite mining by Vedanta Alumina. In 2003, Sterlite Industries, of which Vedanta Alumina is a subsidiary, signed an agreement with the Government of Odisha for a mining project comprising of an aluminium refinery, a captive power plant and active mining in Kalahandi District at an estimated cost of $40 billion (Sahu 2008). In 2004, Vedanta Alumina signed an agreement with the Odisha Mining Corporation to mine bauxite in the Niyamagiri Hills in the Lanjigarh Block of Kalahandi District and the Khambasi Hills of Rayagada District. Both the refinery plant and major mining operations are in Lanjhigarh.

The mining project requires both private and government-owned land as well as a large tract of forest land. Based on the recommendations of two special committees, the central government, in 2010 denied forest and environment clearance to the Vedanta mining project. A four-member committee that visited the area reported:

> The Ministry of Environment and Forests cannot grant clearance for use of forest land for non-forest purposes because the legal conditions for this clearance as laid down by its circular of 3 August 2009, have not yet been met. These include the following: *the process of recognition of rights under the Forest Rights Act has not been completed; the consent of the concerned community has neither been sought nor obtained; and the Gram Sabhas of the area concerned (hamlets in a Scheduled Area has not certified both).*
>
> (Saxena et al. 2010: 8–9, italics original)

The committee further noted:

> Since the Kutia and Dongaria Kondh are heavily dependent on forest produce for their livelihood, this forest cover loss will cause a significant decline in their economic well-being. It must be noted that the Vedanta proposal

> assumes that no displacement will be caused by the mining project whereas there is overwhelming evidence that mining will not only result in widespread resource displacement but may well permanently undermine the survival of the Dongaria Kondh.
>
> (Ibid.: 34)

Without forest and environmental clearance, the mining operation could not start. The refinery, however, has already been constructed. The project has been dominated by controversies and is facing the wrath of Dongaria Kandh, an Adivasi community that has been living for generations in the Niyamgiri Hills where Vedanta has proposed mining (Padel and Das 2010). Kutia Kandha, another Adivasi community residing in the area, is also resisting the mining project.

The Niyamagiri Suraksha Samiti (NSS), an informal community organization, is the vanguard of the resistance to the Vedanta mining project. The movement, with support from a strongly networked Dongaria Kandha, has been protesting to save the Niyamagiri Hills from bauxite mining by Vedanta. The discourse of the movement echoes with that of the PSSP against the UAIL. The movement believes that the Vedanta mining project underestimates the number of people facing displacement. The mining project has also become suspect in the eyes of local residents because the project requires large tracks of forest land. Furthermore, the Niyamgiri resistance has a strong religious overtone because its discourse is not only about protecting the Niyamgiri Hills and forests as livelihood resources and habitat for people, but also about protecting Niyamagiri as a 'sacred land', as Dongaria Kandha believe that their deity 'Niyam' resides in the hills and forests. In the words of Sikoka Budhga:

> We can never leave Niyamgiri. If the mountains are mined, the water will dry up. The crops won't ripen. The medicinal plants will disappear. The air will turn bad. Our gods will be angry. How will we live? We cannot leave Niyamgiri.
>
> (Saxena et al. 2010: 34)

Forest and environment clearance, a mandatory process by the central government to approve the proposed mining project, has become a thorny issue for the Vedanta. The Central Expert Committee of the Supreme Court, the apex court in India, in 2005 recommended the revocation of the Vedanta mining project on the grounds of environmental costs involved. The court put the mining project on hold and asked the company to submit a fresh proposal and to consider the safeguard required. The mining project was denied clearance by the Ministry of Environment and Forest in 2010. In 2013, the Supreme Court recognized the rights of Adivasi and directed that the *gram sabhas* of the villages affected by the Vedanta project hold a referendum to make a decision. The court held that the Constitution of India gives people the right to preserve their sacred land and that the Niyamgiri is sacred to Dongria Kandha.

The *gram sabhas* held in twelve villages unanimously rejected the mining project. During the *gram sabha* in Japara village, an elderly woman said to the presiding panel: 'Niyamgiri is ours; we won't leave the place at any cost. We don't know how to work as a paid labour. If government forcefully displaces us, where then we will go for the work? We won't like to be a bonded labour of anyone, we would prefer to be shot dead but will not leave our Niyamgiri' (Jena 2013: 15). A strong resentment against the government for not providing the basic needs of people finds expression in the movement's narrative. This resentment was articulated in the *gram sabha* of another village: 'Niyamgiri is our hospital; he gives us medicine. We do not have any facility, only now you are coming saying you will give us hospitals. Niyamgiri is our God. We won't give it to anyone. Leave us' (Ibid.). Despite the *gram sabha* verdict and the pending environment clearance, the Vedanta mining project has not withdrawn from the area. Resistance to the mining project has continued as people fear the company will eventually get the necessary permission to continue with the mining operations.

The state is the central authority in facilitating industrial projects. The laws and regulations pertaining to land acquisition, social and environmental impact assessment, environment clearance, and the resettlement and rehabilitation of displaced people on the one hand make the state responsible for the poor, and on the other vest power in the state as the ultimate decision maker. The UAIL has been able to conduct its mining operations because the government allowed the police to supress the protests. The Vedanta mining project has not been given the mandatory mining permission by the central government, but that has not made the provincial government stall the project and force the company to withdraw from the area.

Even the judiciary, while providing a hearing to the protesters and upholding the rights of the Adivasi, has not been able to counter the claims of the mining companies as evident in the case of the Vedanta mining. A decade earlier, in what was considered a landmark verdict of the Supreme Court concerning the claims of the Adivasi against mining operations in another part of India, the court recognized the land rights of the Adivasi yet could not go against the need for mining to propel economic growth. The conflict between the Adivasi economy and mining operations cannot be missed in the observations made by the court:

> The object of the Fifth and Sixth schedules to the Constitution … is not only to prevent acquisition, holding or disposal of the land in Scheduled Areas by the non-tribals from the tribals or alienation of such land among non-tribals inter se but also to ensure that the tribals remain in possession and enjoyment of the lands in Scheduled Areas for their economic empowerment, social status and the dignity of their person. Equally exploitation of mineral resources for national wealth undoubtedly, is for the development of the nation. The competing rights of tribals and the state are required to be adjusted without defeating the rights of either.
>
> (Government of India 1997: 58)

The social contract, the state and the marginalized

Democracy and development are the twin pillars on which the social contract between the state and its citizens was constructed in the wake of India's independence from colonial rule. The anti-colonial movement shaped the aspirations of people. A democratic political regime founded on the liberal principles of rights, a welfare state premised upon special provision for the socially and economically marginalized and a development model based on planned development promised to fulfil those aspirations. A written constitution carefully drafted after laborious deliberation was adopted to guarantee equality in a society where socio-economic hierarchies are deeply entrenched. Economic development combined with social justice thus became the hallmark of the social contract in order to balance the agenda of economic development with the agenda of welfare and redistribution (Mohanty 2018). India adopted a mixed economy to propel development and prosperity together with plans for redistribution and poverty elimination. The state assumed the role of provider, protector and guarantor rights and welfare. Faith on the state was unshakeable. The years since independence have revealed the dichotomies, contradictions, fractures and unfulfilled promises of the social contract. Yet, the belief in a progressive social contract still fills the marginalized with hope and expectation from the state.

The balance between welfare and economic growth has been a challenging task for the state. While India continued to be a welfare state with an agenda of redistribution until the liberalization of the economy in the early 1990s, the resources and benefits of economic development have been appropriated by the landed elite, the industrial class and the upper castes historically placed in a dominant position to appropriate state resources and developmental benefits after independence (Bardhan 1984, 1988; Dhanagre 1987; Kothari 1986; Kohli 1987, 1988). Liberalization ushered in the free market economy. Pushed by pressures from grassroots movements, the state has been trying to find a balance between a liberalized economy based on growth and social welfare. On the one hand, there is the demise of the welfare state and rise of a neoliberal state. On the other hand, there is rights-based legislation to ensure social protection, safety nets and resource rights. The National Rural Employment Guarantee Act 2006, the National Food Security Act 2009 and the Forest Rights Act 2006 are considered landmark pieces of legislation to protect the poor even as the state has consolidated its power in pushing economic growth with an unprecedented ruthlessness. While the attempt to balance the welfare agenda with neoliberal economic growth can be seen as an attempt to reconstruct the social contract to fulfil the aspirations the state had promised in the wake of independence, the protests against large-scale mining are indicators that this balance is tilted in favour of the economic growth agenda.

The historical context and evolution of particular forms of political order gives rise to particular forms of social manifestation of aspiration or discontent. For example, land reforms for the distribution of free land to landless and marginal farmers as a measure to ensure social justice points to a form of socialist and democratic orientation of the political order. A sharp contrast to this is the economic growth agenda through mining. Here the state goes to any extent to divert the livelihood resources of the poor without any prior information or deliberation. It invokes the 'public good' argument that entitles the state to take land for what it considers to be a 'national good'. Resistance to such state action by citizens often invokes suppression and violence from the state. In this context, not only does the state shed its socialist orientation in order to adopt a neoliberal market economy, but its democratic orientation is also compromised.

There is a conflict between the centralization of power and decision making to promote economic growth and the institutions of decentralized governance mandated by the Indian Constitution that vests power and decision making in local governance systems. The Land Acquisition Act is a central government legislation that allows acquisition of public and private land for 'public good' or 'national good'. The Adivasi areas are governed by the decentralized legislation of the PESA that entrusts power onto local institutions to make decisions about the transfer of Adivasi land. Although such institutions are often co-opted or rendered helpless by powerful industrial forces that grab land without involving the local institutions, the Supreme Court direction in 2013 in the case of the Vedanta mining operations in the Niyamagiri Hills in Odisha reinforced village assemblies as a part of the local governance institutions as the ultimate decision-making forum for land acquisition in Adivasi areas. However, despite the village assemblies' decision against the UAIL and Vedanta mining projects, the provincial government did not stall the mining projects of the two companies, thus rendering decentralized governance ineffective.

Another aspect of the state that fractures the social contract between the state and its vulnerable population is violence. While the welfare policies and rights-based legislation promote a state-citizen relation in which the state comes across as patron, well-wisher, rights giver and resource allocator to the poor, this relationship is questioned by people who experience state violence in various forms, among which physical violence features prominently. The agencies of the state such as police, bureaucrats and contractors are ever present in the everyday lives of the poor, and nowhere is the presence of the state more visible than in Adivasi regions where from managing the forests to fighting Maoist insurgency to protecting mining companies, the presence of the state and the violence it creates cannot be missed. Unsurprisingly, the resistance to secure land right against mining companies invites the wrath of the state, even as the state administration implements welfare projects in the same area where the resistance is located.

To understand the significance of the state, it is important to understand the nature of the post-colonial state and the depth of people's relationships of dependence and patronage with it. It is important to understand how the state features in the imagination of people, since it is their relationship with the state, ranging from disillusionment and despair to seeing it as a patron and a benefactor, which is reflected in their relationship with the state. As Chandhoke puts it:

> political preferences for the state over other actors are the outcome of historical processes … that preference formation takes place in a historical context, that of specific institutions or systems of rules. These shape interest, fix responsibility and guide the formation of expectations.
>
> (2005: 1037)

The resourceless people continue to look up to the state to intervene and solve their problems. For historical reasons, the state still looms large in the perception of millions of people.

Poor people's relationship with the state is often complex and ambiguous. The people protesting against mining projects are also the beneficiaries of various welfare and anti-poverty programmes sponsored and implemented by the state. They protest one set of state action, but they are also dependent on another set of state action for patronage and welfare. Their relationship with the state is thus far from perfect – the state is indispensable, but it is also inadequate (Hansen and Stepputat 2001). The state can be authoritarian, but it is also a source of patronage and welfare. How people see the state and how the state comes across to people depends on which agency of the state is closer to them in their daily interactions (Corbridge et al. 2005): it could be the Border Security Force or a Block Development Officer. People thus have multiple images of the state: the same state that enacts the right to food security for the poor also takes away their land for mining operations. The giver of rights becomes the destroyer of livelihoods. The state, with its institutional settings, programmes, discourses and strategies, has multiple facets vis-à-vis citizens. From policy formulations that take place at the highest level of the state to their implementation through a vast network of bureaucracy from the district level downwards, the manifestations of the state are diverse and at times in conflict with each other. The aspect of the state that captures the imagination of people is often determined by contextual factors. A welfare state provides social schemes such as employment guarantee and social security, developmental goods such as housing and water, public education and health care as well as a host of other special welfare measures to the Adivasi. The state in this context comes across as a supportive state. Likewise, when the state listens to the grassroots demands for forest rights, it is seen as an enabling state. The same state is also seen as a hostile state when it diverts Adivasi land for mining and supresses protests.

The state is at once a distant impersonal idea as well as a localized and personified institution (Gupta 2001). The distant idea gives rise to the ideal state people would like to have, but the personified institutions of the state are often a source of anxiety. Since post-colonial states came into power after prolonged periods of radical nationalist struggle, their origin makes them benevolent in the eyes of their populace. However, as post-colonial states continue to fashion themselves after the power structures and motives of the colonial state, their actual behaviour is often different from how they are imagined by the poor. People straddle between an ideal state as manifested in the Constitution, legislations and the role the state plays as rights-giver and protector and the actual state they come across. Kalahandi, where the Adivasi have been protesting against the Vedanta mining project, has witnessed the extremes of state negligence: in the successive droughts that led to hunger and death, the state failed to provide help; instead it gave a spin to the dire situation by emphasizing that there were no deaths due to starvation (Jayal 2001). People, however, still believe that the state will listen, that they can pressure the institutions of the state to withdraw the mining project.

The promises of a post-colonial democratic and developmental state on which the social contract was founded have been compromised. The social contract has suffered fractures and has witnessed a gradual erosion of the values that had guided the state in the wake of independence from colonial rule. However, the promises have not been completely abandoned by the state as evident in the welfare and social protection measures that have been put in place as well as the policies and legislations that protect the land and forest rights of the Adivasi. People have come to experience many facets of the state that have shaped their perception of the state and expectation from the state at different moments in time. The Adivasi protests' articulation of rights and claims on the state indicate that the promises of the social contract still animate their imagination.

Conclusion

The conflict between the Adivasi economy that views land as a means of livelihood and sustenance and the mining projects that want to use land for profit raises the issues of ownership of and control over subsistence resources of the poor. Even though the governance of Adivasi land secures such rights in favour of people, the state as the promoter of economic growth overrules its own promises. The protests articulate the resource rights and resource justice for the poor Adivasi.

The retreat of the state under neoliberal arrangements, its new role as a mere facilitator and the midwife role it performs for economic growth have resulted in the loss of its power to protect the rights of its vulnerable citizens. It is not merely that redistribution of benefits of economic growth has not accrued to Adivasi whose land was taken away; the normative role the state plays in protecting the rights of its citizens has also suffered. The state does not respond to protests and remains indifferent, or it responds partially and at times brutally. The protests are sites where the fractures in the social contact between the state and people are palpable. They are also sites where the negotiation of the social contract and expectation from the state animates Adivasi imagination and action. The protests are an indication of why the state still matters to its most vulnerable citizens.

Notes

1 The term Adivasi connotes a collective of heterogeneous aboriginal minority communities in India called tribals. Adivasi literally means 'original inhabitants of the land'.
2 The Indian planning system was based on five-year plans made by the Planning Commission. The First Five Year Plan was for the period 1951–1956. The five-year planning continued until 2017.
3 On 2 January 2006, in the Special Economic Zone of Kalinaganagar in Odisha, twelve Adivasi were shot dead by police for opposing the mining operations of Tata Steel Corporation, an Indian private-sector company.
4 Panchayat is a system of rural local governance mandated by the Indian Constitution. The governance system is comprised of *gram sabha* (the village assembly of all adult members residing in the cluster of villages that come under a panchayat as well as an executive council of representatives elected by *gram sabha* members.
5 Most big companies operating in Adivasi areas have their own campuses to provide accommodation, school and health care facilities to their employees.
6 A block is a middle-level administrative unit of a three-tier decentralized governance that begins at village level and ends at district level. A block comprises of a set of revenue villages, and a district comprises of a set of blocks.

7 The Fifth and Sixth Schedule of the Indian Constitution contains special administrative arrangements for Schedules Areas, areas demarcated as having a large Adivasi population.

References

Agarwal, A. 1985. 'Ecological Destruction and the Emerging Patterns of Poverty and People's Protest in Rural India', *Social Action* 35(1): 54–80.

Ambagudia, J. 2010. 'Tribal Rights, Dispossession and the State in Orissa', *Economic and Political Weekly* 45(33): 60–67.

Baboo, B. 1991. 'Big Dams and the Tribals: The Case of the Hirakud Dam Oustees', *Social Action* 41(3): 288–303.

Bandopadhyay, J., and V. Shiva. 1988. 'Political Economy of Ecology Movements', *Economic and Political Weekly* 23(24): 1223–1232.

Bardhan, P. 1984. *The Political Economy of Development in India*. New Delhi: Oxford University Press.

Bardhan, P. 1988. 'Dominant Proprietary Classes and India's Democracy', in A. Kohli (ed.), *India's Democracy: An Analysis of Changing State-Society Relations*. New Jersey: Princeton University Press.

Bavisker, A. 1995. *In the Belly of the River: Tribal Conflicts over Development in the Narmada Valley*. New Delhi: Oxford University Press.

Bhatt, C.P. 1991. 'Chipko Movement: The Hug that Saves', *Survey of the Environment*. Chennai: The Hindu (newspaper).

Chandhoke, N. 2005. 'Seeing the State in India', *Economic and Political Weekly* 40(11): 1033–1039.

Corbridge, S., G. Williams, M. Srivastava and R. Veron. 2005. *Seeing the State: Governance and Governmentality in India*. Cambridge: Cambridge University Press.

Das, A., and Das, V. 2006. *Chronicles of a Struggle and Other Writings*. Kashipur: Agragamee Publications.

Dhanagre, D.N. 1987. 'Green Revolution and Social Inequalities in Rural India', *Economic and Political Weekly* 22(19–21): AN137–AN144.

Fernandes, W. 1991. 'Power and Powerlessness: Development Projects and Displacement of Tribals', *Social Action* 41(3): 243–270.

Gadgil, M., and R. Guha. 1994. 'Ecological Conflicts and the Environmental Movements in India', *Development and Change* 25(1): 101–136.

Goodland, R.J.A. 2007. *India: Orissa, Kashipur: Utkal Buaxite and Alumina Projects: Human Rights and Environmental Impacts*. Retrieved 30 June 2020. https://www.business-humanrights.org/sites/default/files/media/bhr/files/Goodland-Utkal-Mar-2007.pdf.

Government of India. 1997. *Judgement in the Case of Samatha versus the State of Andhra Pradesh and Others*. New Delhi: Supreme Court of India.

Government of India. 2011. 'Census of India', *Office of the Registrar General & Census Commissioner, India*. New Delhi: Government of India.

Guha, R. 1989. *The Unquiet Woods: Ecological Change and Peasant Resistance in the Himalaya*. New Delhi: Oxford University Press.

Guha, R., and M. Gadgil. 1995. *Ecology and Economy: The Use and Abuse of Nature in Contemporary India*. New Delhi: Penguin.

Gupta, A. 2001. 'Governing Population: The Integrated Child Development Services Program in India', in T.M. Hansen and F. Stepputat (eds.), *States of Imagination: Ethnographic Explorations of the Postcolonial State*. Durham: Duke University Press.

Hansen, T.M., and F. Stepputat. 2001. *States of Imagination: Ethnographic Explorations of the Postcolonial State*. Durham: Duke University Press.

Jayal, N.G. 2001. *Democracy and the State: Welfare, Secularism and Development in Contemporary India*. New Delhi: Oxford University Press.

Jena, M. 2013. 'Voices from Niyamgiri', *Economic and Political Weekly* 48(36): 14–16.

Kohli, A. 1987. *The State and Poverty in India: The Politics of Reform*. Cambridge: Cambridge University Press.

Kohli, A. 1988. *India's Democracy: An Analysis of Changing State-Society Relations*. New Jersey: Princeton University Press.

Kothari, R. 1986. 'Masses, Classes and the State', *Economic and Political Weekly* 21(5): 210–216.

Ministry of Law and Justice. 2007. 'The Scheduled Tribes and Other Traditional Forest Dwellers (Recognition of Rights) Act, 2006'. New Delhi: Ministry of Law and Justice, Government of India.

Ministry of Law and Justice. 2013. 'The Right to Fair Compensation and Transparency in Land Acquisition, Rehabilitation and Resettlement Act, 2013'. New Delhi: Ministry of Law and Justice, Government of India.

Ministry of Tribal Affairs. 1996. 'The Provisions of Panchayats (Extension to Scheduled Areas) Act, 1996'. New Delhi: Ministry of Tribal Affairs, Government of India.

Mishra, S.K. 2002. 'Development, Displacement and Rehabilitation of Tribal People: A Case Study of Orissa', *Journal of Social Sciences* 6(3): 197–208.

Mohanty, B.B. 1997. 'State and Tribal Relationship in Orissa', *Indian Anthropologist* 27(1): 1–17.

Mohanty, B.B. 2005. 'Displacement and Rehabilitation of Tribals', *Economic and Political Weekly* 40(3): 1318–1320.

Mohanty, R. 2003. 'Save the Chilika Movement: Interrogating the State and Market', in R. Tandon and R. Mohanty (eds.), *Does Civil Society Matter? Governance in Contemporary India*. New Delhi: SAGE Publications.

Mohanty, R. 2018. *Democratization of Development: Struggles for Rights and Social Justice in India*. New Delhi: SAGE Publications.

Padel, F., and S. Das. 2010. *Out of This Earth: East India Adivasis and the Aluminium Cartel*. New Delhi: Orient Blackswan.

Pal, S. 2020. 'Odisha Anti-Mining Protests: 40 Women, 12 Children Languish in Rayagada Jail, Say Locals'. *Newsclick*. Last accessed 30 June 2020. https://www.newsclick.in/Odisha-Anti-mining-Protests-40-Women-12-Children-Languish-Rayagada-Jail-Locals.

Pathak, A. 1994. *Contested Domains: The State, Peasants and Forests in Contemporary India*. New Delhi: SAGE Publications.

Pati, B. 2013. 'The Diversities of Tribal Resistance in Colonial Orissa, 1840s–1890s: Survival, Interrogation and Contests', *Economic and Political Weekly* 48(3): 49–58.

Patibandla, S. 2009. 'Tribal Movement in Orissa: A Struggle against Modernisation?' *Working Paper 215*. Bangalore: Institute for Social and Economic Change.

Planning Commission. 2001. 'Approach Paper to the Tenth Five Year Plan (2002–2007)'. New Delhi: Planning Commission, Government of India.

Ray, S., and S. Saini. 2011. 'Development and Displacement: The Case of an Opencast Coal Mining Project in Orissa', *Sociological Bulletin* 60(1): 45–64.

Sahoo, M. 2015. 'Mining and Land Acquisition: An Analysis of Mineral Rich Tribal Regions in India', *Third World Studies* 32(2): 153–174.

Sahu, G. 2008. 'Mining in the Niyamgiri Hills and Tribal Rights', *Economic and Political Weekly* 43(15): 19–21.

Sarangi, D., S. Pradhan and S. Mohanty. 2005. 'State Repression in Kashipur', *Economic and Political Weekly* 40(13): 1312–1314.

Satpathy, B. 2017. 'Forest Rights Act Implementation in Odisha: Redressing Historical Injustices', *South Asia Research* 37(3): 259–276.

Saxena, N.C., S. Parasuraman, P. Kant and A. Baviskar. 2010. *Report of the Four Member Committee for Investigation into the Proposal Submitted by the Orissa Mining Corporation for Bauxite Mining in Niyamagiri*. New Delhi: Ministry of Environment and Forest, Government of India.

Stanley, W. 1996. 'Machkund, Upper Kolab and NALCO Projects in Koraput District, Orissa', *Economic and Political Weekly* 21(24): 1530–1538.

8 Claiming agency by telling a counter-story in court

Adivasis v. ‘encounter’ killings in India

Shylashri Shankar

Introduction[1]

It is often alleged that legal narratives are structured to silence, exclude and oppress those who are not part of the dominant culture. The behaviour of the Supreme Court of India during the COVID-19 lockdown seems to confirm such an allegation.[2] To a plea asking it to direct the central government to help thousands of migrant workers stranded in cities and towns and needing money, food, shelter, water and medical services, the judges expressed their helplessness. ‘Trust us’, said the central government of India. Don’t trust them, was the implicit message in the petition. ‘The affidavit by the Solicitor General claims that everything is being done. But then, the order has not so far been implemented on the ground’. ‘What are we to do?’ said the judges, adding that the judiciary could not run the government.

Meanwhile, a thousand migrants assembled in Prime Minister Modi’s home state of Gujarat and demanded to be allowed to go home. The police arrested seventy and used sticks to disperse the rest. These migrants from rural areas provided cleaning, cooking and other services to the middle and upper classes living in India’s cities. They had built malls, roads, restaurants and apartment buildings. Among these migrants who live on daily wages are the scheduled castes (those at the bottom of Hinduism’s caste hierarchy) and the scheduled tribes (indigenous people) who, together, make up a quarter (325 million) of India’s population of 1.3 billion people.[3]

What would the migrants conclude about how they fare in India’s democracy? They would conclude that their lot is one of being invisible to the state, excluded from constitutionally guaranteed rights and protections, and of being silenced if they raised their voices. Critical legal studies show how the legal setting silences subaltern voices, how their dreams and desires are omitted by the lawyers arguing their case (Parmar 2015), and how the law is biased against the marginalized (Delgado 1989).

However, not all stories told in court by such groups are doomed. Some stories can and do challenge the dominant narrative successfully. Counter-stories can transform the legal system into more inclusive and responsive frameworks for these ‘outsider’ groups, argues Richard Delgado (1989).[4] Telling these stories, he says, allows the outsider group to heal, creates group solidarity (the sense that ‘I am not alone’) and forces the oppressors to see themselves as such.

Following Delgado's rationale, this chapter argues that the encounter between marginalized citizens and the state in a court is an important aspect in creating a sense of citizenship for these groups. While creating narratives for the legal process, the marginalized and the powerful are pushed to calibrate their world views, or their 'nomos' in Robert Cover's (1983) words, to influence the outcome. This is because law, as Cover points out, is better seen as a normative world, a nomos in which legal rules and institutions interact with other cultural forces to produce legal meaning.

> We inhabit a nomos – a normative universe. We constantly create and maintain a world of right and wrong, of lawful and unlawful, of valid and void ... No set of legal institutions or prescriptions exists apart from the narratives that locate it and give it meaning. For every constitution there is an epic, for every decalogue, a scripture. Once understood in the context of the narratives that give it meaning, law becomes not merely a system of rules to be observed, but a world in which we live.
>
> (Cover 1983: 16)[5]

A speaker tells a listener what someone did to get what he/she wanted and why – that is the simple definition of a story in John Truby's *The Anatomy of a Story* (2008). In a legal process, this definition is split up into parts, and each role is assigned to a different character. A speaker (the litigants) tells a listener (the police, lawyers and, finally, the judge) what someone did to get what he/she wanted (the plaintiff and the defendant tell different accounts here). Figuring out the 'and why' part is assigned to the judge, not to the litigants. The judge has to sift through the parochial perspectives and create a representative view of guilt and innocence that will allow the law to do something about the suffering.

When we approach courts with our stories about being wronged, there is, as Joanna Wheeler's chapter in this book suggests, a distinction between stories told at a personal level and by a collective. Wheeler claims that: 'The first is about surfacing and claiming a position; the second, in the collective process, is about articulating what forms of justice are required in relation to that trauma. ... In the second layer, the emphasis shifts from the difficult task of articulating experiences, especially traumatic ones, towards making sense of these experiences in political terms'.

To make sense of an experience in political terms in a democracy implies telling the counter-story in the language of rights. This chapter analyses a counter-story told by two young women who are Adivasi (the literal meaning of Adivasi is 'forest dweller'), or part of a 'scheduled tribe' in India's legal language.[6] They challenge the story of the paramilitary and police personnel who say that their actions were against villagers suspected of mounting armed resistance to their authority. The chapter charts their counter-story's passage from a personal level (individuals in several villages experiencing trauma at the hands of the security forces) to the shaping and telling of a collective story in court. By focusing on the story (or stories) told in the petition, it explores how engaging with the narrative framework of the law assists the villagers to make sense of these experiences in political terms and demand justice within the framework of constitutional rights.

Using the tools of storytelling, namely the appeal to the affective, persuasive and transformative aspects of the cognitive process highlighted by Winter (1989), the analysis assesses the following in the petition[7]:

- How the narrative tackles the formal elements of legal storytelling where context is erased and 'facts' are emphasized.
- How the petition builds bridges of understanding between two nomos(es) – the Adivasi and the non-Adivasi (judges, lawyers and others).
- How the vision of what the Adivasi expect from the Indian state is shaped by their engagement with the legal process, and how their vision shapes the dominant narrative.

The method uses a stylized portrayal of these elements in a single case. The chapter does not make the claim that the lawyers consciously followed these elements in writing the petition, nor does it claim to portray the thoughts of the Adivasi or the judges.

The petition

Bastar in central India is considered by the Indian state as the site of Maoist Naxalite activity. Over 70% of those who live in Bastar are Adivasi.[8] Their villages experience frequent raids by paramilitary forces who arrest and kill suspected Maoists in 'encounters', a euphemism for state-sanctioned extra-judicial killings.[9] These 'encounters' occur when teams of local police and paramilitary forces go out on combing operations after receiving 'verified information' about the presence of Maoists in the area.

Suneeta Pottam (age 19) and Munni Pottam (age 18), residents of Korcholi village, the site of one of these 'encounters', filed a Public Interest Litigation (PIL) along with a WSS (Women Against Sexual Violence and State Repression) at the Chattisgarh High Court. The PIL challenged the extra-judicial executions by paramilitary forces of six people in Kadenar, Palnar, Korcholi and Andri in Bijapur district in 2016. In 2019, their petition was transferred to the Supreme Court of India, which was hearing similar cases.[10]

The petition contained the eyewitness accounts of villagers and the families of the deceased. In one village, a married couple were forced out of their home in the evening at gunpoint, on the pretext of getting them 'surrendered' at the Gangaloor police station. Their bodies were later delivered to their family.[11] In another village, a man was dragged away from the fields that he was ploughing, with his hands tied behind him and in full view of his young wife, mother and other villagers. In the third village, some women saw a man shot by the police while he was visiting his relatives' house.[12] The police said that these people were among those who had been shot in the 'encounters'.

When the villagers sought help from the police, they were rebuffed. Sodhi Hurra, the father of 9-year-old Sodhi Sunna, said the Bijapur police did not even allow him inside the police station when he went to report his missing child. Sukli Hemla, the elderly mother of Seetu Hemla, went to the Gangaloor police immediately after Seetu had been captured and dragged into the jungle. However, the police did not lift a finger to help her, she said.

Let us turn to how the petition forged a counter-story from these powerful allegations against the police and paramilitary.

Telling stories in court

'Judges are people of violence; judges are people of peace', said Robert Cover (1983) in *Nomos and Narrative*. 'Because of the violence they command, judges characteristically do not create law, but kill it, that is, they kill the other legal traditions blooming from the nomos of different groups by asserting "this one is the law"' (Cover 1983: 53). Cover points out that judges use the concept of jurisdiction to legitimize the violence of the state towards those who do not agree with their interpretation. We saw an instance of such use by the Supreme Court of India in the petition on the migrants' plight during COVID-19. But judges are people of peace too, as Cover states. They assert a regulative function, one that permits a life of law rather than violence; but this occurs only after the community acquiesces to the judge's interpretation.

Suneeta and Munni's counter-story faces a difficult challenge because law in liberal, illiberal and post-colonial democracies continues to be used by the powerful to exclude the voices of the powerless. The privileged, as Scheppele (1989) points out, live in a reality where their self-believed stories are officially approved and become facts, while the self-believed narratives of the non-privileged are officially distrusted, rejected, found to be untrue or not heard at all. Such situations in a democracy raise fundamental issues of legitimacy for the rule of law. "Applying models of consent (Locke's (1689) social contract or Rawls's (1971) veil of ignorance) to abstract laws or to a legal system does not help us tackle the question of legitimacy because these models do not require that people's particular points of view be taken into account at all," says Scheppele (1989). Consent, she says comes apart in battles of description; consent comes apart over whose stories to tell. The experiences and cultural tropes of the Adivasi (who are seen as outsiders) are alien to the in-group, who include the lawyers and the judges.

In a symposium on legal story-telling published in the *Michigan Law Review* three decades ago, legal scholars mulled over the following question: How can we include the perspectives, dreams and desires of the marginalized in law's construction of an official story for a particular case? They problematized law's workings in the context of how stories are allowed to be told in court, and the result of this on the marginalized – especially when the lawyers and judges have a different cultural background and cognitive predilections.

The legal process

Adivasi in India and indigenous people in other countries face several problems in the legal process. This section uncovers how the rules of the legal process play out for Adivasi, and shows how misaligned the assumptions of participatory democracy are about neutrality, fairness and justice, and how post-colonial forms of politics replicate colonial power structures.

In the common law system followed by India (and the other former British colonies), guilt is determined by establishing that the defendant committed the crime. However, the rules of evidence forbid presenting evidence of prior crimes unless it is relevant to issues other than the defendant's character (Posner 2009). To protect the defendant, the prosecution is forbidden from telling the story with the richness of detail used in fictional narratives.[13] The legal process thus mangles or completely silences the telling of the story in several ways (Cooke 2002; Eades 2013; Parmar 2015), including:

(i) *Emphasis on structuring in short bits.* In an adversarial system, lawyers for both sides want the other side to stick to a narrative that responds to the 'What happened next?' question; they do not want the other side to tell a story about why something happened. Outsiders' stories are often excluded by the daily operation of apparently harmless legal habits such as beginning the story when the trouble began rather than a 'wide angle' version which puts the event in a broader context (Scheppele 1989). The interviewer's questions decide the sequence of events, the information that can be told, the details that need to be withheld and the order in which this information is told. This usually upsets the storyteller's rhythm. Additionally, injunctions like 'Look in front' or 'Speak louder' unnerve the witness.

(ii) *Devaluing variations and frankness.* Variations when a story is re-told are not viewed positively by the court's rules and are usually treated as lies. But they need not be. Linguistic research and our own experience shows that we re-tell stories with minor variations. Psychological research reveals that the memory process involves three stages – acquisition, retention and retrieval – and that recollection is a reconstructive process (Parkin et al. 1999). Inconsistencies, therefore, can flow from the way our memory works, and from the social process of reconstructing the narrative. Procedural rules also do not allow the person to reflect frankly on the links between the events. Some norms, such as the lawyer advising her client to remain silent, may deepen the court's suspicions.

(iii) *Cultural dissonance.* Some actions may be interpreted differently because of cultural dissonances between the court's mindset and that of the indigenous citizen. Silence can be interpreted negatively, though it may be a cultural way of thinking about a question.[14] An Australian Aboriginal would say yes to affirm the negative, whereas an English speaker would say no (Cooke 2002).[15] A lawyer narrated an incident to Parmar (2015) where an Adivasi was asked to fix his thumbprint on a testimony. He refused because he did not trust officials. Instead of explaining that it was part of legal procedure, the judge saw it as a wilful flouting of the state's rules and threatened to send him to prison.

Such dissonance extends to the way an Adivasi is imagined by the metropolitan elites. In Parmar's (2015) conversations with lawyers and judges in Delhi, one of them asked if the protestors were 'real' Adivasi – short and dark. He did not even know how to imagine them. Backward, illiterate, marginalized, poor and dumb are the clichés applied to Adivasi in India. The word 'tribal' itself connotes primitiveness. This shows a larger dissonance at work in the legal process – a dissonance between the imaginary of the Adivasi and their expectations from the state, and the dominant imaginary of what the nation ought to be held by judges and even their own lawyers.[16] For instance, state law in India does not recognize tribal rights to maintain connections to their traditional lands in perpetuity.[17] When such rights are trampled in the name of development (e.g. mining and other activities in forests and tribal lands), the lawyers for the Adivasi avoid basing their arguments on tribal rights, especially when such rights seem to diverge from the imagined future of the nation (Parmar 2015). The legal process that is supposed to enable them to speak is put to work in this eclipsing.

Judges are enmeshed in structures of social meaning, a nomos which they share with the other officials of the state, including the police. As Cover (1983) points out, judges have a jurispathic tendency to favour the state's point of view. The stories told by the dominant group, in this case the police, 'remind it of its identity in relation to the outgroup, and provide it with a form of shared reality in which its own superior position is seen as natural' (Delgado 1989: 11). The outgroup, in this case the Adivasi, create their own story, which fosters a shared understanding and cohesion among them and 'circulates within the group as a kind of counter-reality' (Delgado 1989: 11).

The standard narrative of law as being neutral and judges as being impartial is not played out in reality, where interpretations often favour the dominant nomos. The counter-narrative has to push the judges and lawyers away from seeing the world in a binary way – our culture versus the Adivasi culture, civilized versus jungle folk – and encourage them to understand other ways of thinking. Not just that, the petition must woo them to its side. Munni and Suneeta's petition faces a formidable set of barriers. How to make the outgroup's story comprehensible to the dominant nomos and how to tilt the scales in the outgroup's favour – that is the challenge for Munni and Suneeta's petition.

The process of creating a counter-story is an important part of how the Adivasi make sense of their traumatic experiences at a political level. Meaning is located in the process through which the different actors interact with the world. As Das and Poole (2004) say, any question of justice and rights must arise not from the moral space of innocent victimhood, but from the rough and tumble of everyday life. To identify and link the harm conceptually with a larger narrative of rights, the Adivasi have to engage with the gritty process of registering a complaint, finding lawyers and translating their imaginaries into a co-created counter-story.

What are the cognitive elements that shape a powerful counter-story in the legal arena? Let us examine how Munni and Suneeta's petition forges affectiveness, persuasion and, finally, transformation – the three elements needed to tell a powerful counter-story.

The art of telling a counter-story in court

For a counter-story to be successful in court, it has to establish representativeness, unreflexivity and reliability so that a prevailing order can credibly justify itself through law (Winter 1989). As he explains, this means that the law must be able to cover a wide variety of situations to which it can be applied (i.e. generalizability). However, such 'trans-situational applicability' (Parmar 2015: 97) does not occur for the Adivasi because the issue is often transformed in such a way that it eclipses their voices.[18] The second requirement of unreflexivity is also difficult for the counter-story to achieve. For the law to appear legitimate, it must evoke an automatic sense of validity that is only provided by automatic cultural knowledge (Winter 1989). However, no such shared cultural knowledge exists between the Adivasi and the non-Adivasi. This also impacts representativeness because for a law to function effectively and equitably, it must be communicated with as high a degree of relative reliability as possible.

How can the counter-story circumvent these obstacles? Drawing on Winter (1989: 2272), let us see how the petition applies three elements of the cognitive process that determines how a person makes sense of a story. 'First, the cognitive process is fundamentally imaginative rather than mechanical, that is, we understand and construct our world by imaginative connections through metaphor, metonymy, and the imaginative extension of idealized cognitive models (ICM)' (Winter 1989: 2272). Idealized cognitive models (ICMs) are the stock stories that structure and make the regular aspects of our daily experiences meaningful. Second, he says, we share an ICM of narrative that depends upon and encodes a particular way of making sense of the world. Third, 'though ICMs are grounded in experience, they are not objective representations of that experience but idealized more-or-less characterizations' (Winter 1989: 2272). These three points, he says, are jointly the fulcra of the success of the three types of prescriptive narratives – affective, persuasive and transformative storytelling.

Affectiveness

Affectiveness is produced when the reader imagines herself in the shoes of the protagonist in the story. Usually we are able to do so, says Winter (1989), because the ICM of narrative is structured by a source-path-goal schema and grounded in our common experience of journeys. The judge has to imagine himself in the shoes of the protagonist. The question is, who will the judge choose as the protagonist – the police or the Adivasi? For the police and security agencies accused in the petition, the task of making the judge identify with them is easier because they share cultural assumptions about the Adivasi and also belong to the same network of institutions within the Indian state. It is hard for a judge in India, who tends to be male, upper caste, upper class and urban (Shankar 2009) to imagine himself as a forest dweller in an area where Maoist rebels are operating. It is also hard for him to imagine what it is like for the Adivasi to fear violence from the very authorities (including judges) who are supposed to protect them.

The petition therefore has to invoke an existing storyline or ICM that would be familiar to the judge, but at the same time, in doing so, it must not eclipse the voice of the Adivasi (as we saw earlier in the forest rights cases). It does so in three ways.

By Connecting the petitioners with the qualities of a public-spirited citizen

In a traditional court case, the petitioner (as the storyteller) has to have been directly harmed by the action. Public Interest Litigation, a tool created by the Supreme Court of India in the early 1980s, allows a concerned citizen to approach the court on matters of public interest. Unlike the regular legal process, it is non-adversarial, and the emphasis is on collaborative fact-finding to reach a solution. Petitions can be sent directly by citizens or by NGOs and activists on behalf of citizens, or even suo moto. The petitioners used the latter method, emphasizing that while they did not have a direct relationship with those who were killed (thus showing them as neutral observers), they had experienced such violence in their lives. The petition says that Munni and Suneeta 'have themselves experienced an extra-judicial execution carried out in their village of an unarmed farmer'. It notes that as children, Suneeta and Munni had to give up their studies when their school closed down due to the violence of Salwa Judum (a state-supported, anti-Maoist civilian vigilante force) in 2005. The vigilantes burnt their houses. With their families reduced to penury, the girls began to work in stone quarries as coolies to support them. The petition praises them as public-minded individuals who, being young and energetic and Hindi-speaking, help their fellow villagers by often representing them before authorities.

The WSS is portrayed as a responsible and committed network of women's organizations in India that takes up issues of sexual violence against women and structural forms of repression. Its legitimacy is established by narrating previous fact-finding missions into mass sexual violence on Adivasi women by police and members of the CRPF (Central Reserve Police Force) in one of the villages. To add heft, the description includes a statement that its previous work on uncovering sexual violence perpetrated by police and the CRPF on Adivasi women in one of the villages was corroborated by the government's National Commission of Scheduled Castes and Scheduled Tribes.

By connecting the victims with the qualities of a good citizen

At its basic level, the petition is about murders. As one lawyer pithily put it, 'in every murder case, there are only two issues: Did the deceased deserve to die? And if so, was the defendant the person to do it?' (Winter 1989: 2272).

The counter-story has to emphasize that the victims did not deserve to die. The petition highlights qualities – hardworking, made mistakes but reformed their ways – that the dominant ICM associates with good citizens.

Manoj Hapka was a hardworking farmer who had gathered enough Mahua flowers this year to have bought two cows. Sukku Kunjam and his brother, Soma, were working in their field when the security forces appeared: 'Fearing arbitrary punishments, arrests and beatings, the two brothers, along with other able-bodied men of Itavar village fled their village'. Another victim, Seetu Hemla, was ploughing his field in the morning 'when the security forces suddenly emerged from the surrounding jungles, encircled him and caught him. They tied his hands behind him, and dragged him away to the nearby jungles, in full view of his family members and other villagers'. Kuhdami Ganga had gone to the jungle with two other young men to collect siyadi leaves in preparation for a wedding the next day. They realized the police were hiding nearby, and fearing arbitrary arrests, they ran back to the village. The police opened fire and killed Kuhdami.

The concept of making mistakes but learning from them and turning around their lives is another familiar theme for the dominant ICM. The petition uses it for two victims, a husband and wife (Manoj Hapka and Tati Pande). They were former Maoists but left the movement after they met and fell in love. They were taken into custody by the security police. The next morning, their bodies were returned to their family 'devoid of any clothing, and with gunshot wounds as well as stab wounds and evidently broken bones'. The villagers later found the shirt worn by Manoj when he was taken. It was hanging from a Mahua tree in the jungle and there were bloodstains below it. Pande's clothes were lying nearby.

Using familiar themes of family, love and grief to create resonance with the Adivasi's plight

In several instances, the families tried to inform the police that the killings were in cold blood, 'but their complaint was not taken with any level of seriousness and not put down in writing'. Seetu Hemla's family

> recovered the body of Seetu the next day from the Bijapur District Hospital. The body seemed to have several gunshot wounds and was also badly mutilated, with flesh wounds, broken bones in the legs and broken finger bones, suggesting that he was subjected to inhuman ante-mortem torture. The family, although illiterate and living in a remote village, has tried to seek justice for their slain son, but with little success.

The belief that children must be protected and must not be harmed links the ICMs of the judge and the Adivasi. The petition highlights the case of a missing child and the suspicious behaviour of the police. A young child, Sodhi Sannu, aged 9 years old, was working in the tomato fields close to his house when the security forces returned to the village. No one in the village witnessed what happened next, but there was the sound of gunfire. Alarmed, the mother and a neighbour rushed to the fields. They saw blood spots on the way, but were prevented from going further by policemen, who threatened to kill them. Sodhi Sannu has not been seen again.

After affectiveness is forged between the judge's ICM and that of the Adivasi at the level of individual stories, the next step in the cognitive process is to show that the harms are not simply one-offs but actually represent a type.

Persuasion

Winter (1989) uses Karl Llewellyn's concept of 'situation-sense', a concept identified with the process of categorization, to distinguish between facts of the individual case and facts of the situation taken as a type. If we apply this distinction to the petition, the facts surrounding the deaths of individual villagers will be the individual cases; while the failure to register the complaints of the victims' families, the presence of members of citizen-militias and 'former Naxalites' in the police operations, and the extra-judicial killings by the security forces will belong to a type. Persuasion involves creating these 'types' for the judge. The petition offers concrete facts and experiences that make the court's idealized cognitive model inapplicable and another ICM more obviously so.[19]

Highlighting factual inaccuracies in the opponent's response

Suneeta and Munni's petition highlights facts that are verifiable and point to a different way of making sense of the events. In Tati Pande and Manoj Hapka's case, the police said the two were killed in an 'encounter'. The petition points out several factual inaccuracies. First, the whole village witnessed their arrest in the village: 'Both the deceased were dressed in simple village clothes, when they were taken from home – Pande was wearing a skirt, blouse and a chunni, and Manoj was wearing a lungi and shirt'. Second, the photographs taken by the police after the alleged 'encounter' show two bodies wearing new uniforms that contain neither blood nor gunshot wounds. The petition notes that 'there is a very low possibility that in an intense gun-battle with 30–35 armed Maoists, 2 people who are killed at random turn out to be a husband-wife couple'. It then raises doubts about the couple's presence so far away.

> The areas of operation of these alleged Maoists, the Darbha Valley and Kanger Valley area are at least 150 kms away from the place where the couple are claimed to have been killed, and no explanation is forthcoming as to what they were doing, armed and uniformed, in an area that is so distant from their normal area of operations.

Third, the petition notes that Tati Pande, the wife, had gone to Bijapur to receive a bonus for gathering tendu patta leaves on the day prior to the so-called 'encounter'. By pointing to these inconsistencies in the police narrative, the petition creates a different and more coherent version of the facts.

Juxtaposing facts in a way that creates links to the context

A skilful narrative can bring in the context, in this case past enmities, without seeming to do so. The petition does this by repeating verifiable statements from eyewitnesses that link the assailants to 'surrendered Naxalites' who either live in the same village or nearby villages and know the victims: 'Tati Bandi and her son, Tati Ramlu, who are both eyewitnesses to this incident, recognize some of the security personnel who entered their house as surrendered Naxalites'. It identifies the personnel as cadre of the CPI (Maoist) party who had frequently visited their village. Similarly,

> Pido Pottam goes on to identify some of the men who had come to the village along with the security forces – they are Sagar Hemla from Palnar, Teera

> Sannu from Pusnar, Manish from Kamkanar, Sallu Bhogam from Palnar and Mangesh Kunjam from Nendra – these are all men from surrounding villages who used to work with the Naxalites and had come to their village earlier. Now they all work for the police.

The conclusion that the petition wants the judge to arrive at is this: that these surrendered militants 'have many of their own reasons for settling scores with these villagers'.

Reiterating common elements in the individual stories

The petition's narrative highlights a pattern of police behaviour: depriving Adivasi of their fundamental right to life without going through a procedure established by law. This is done by pointing out common elements linking the individual stories:

- In several cases, the eyewitness accounts of the villagers challenge the factual elements of the police's story.
- Not just the victim, but also other able-bodied men in the village flee from the police.
- The police employed 'surrendered Naxalites' to flush out their former comrades.
- The police's refusal to register the cases casts doubt on their innocence.
- The conclusions of a fact-finding mission by activists, political party leaders and former legislators confirm the villagers' accusations.
- The addition of other cases with similar allegations to the original petition. These included the extra-judicial execution of Mangu Korme of Peddakorma village, who was captured by the security forces while chasing a monkey that had attacked the village's paddy stocks. He was paraded in the villages as a captive and then killed in cold blood. There was also an attempted extra-judicial killing on Mangu Tati of Palnar village, who was captured by the forces while in the forest trying to gather some bamboos, but he was rescued by the village women who covered his body with their own and would not let the police shoot him.[20]

Tone and non-coercion

To be effective, stories must appear to be non-coercive, says Richard Delgado (1989). They should invite the reader to suspend judgement, listen for their point or message, and then decide what measure of truth they contain. Here, the use of a PIL allows for a non-adversarial and more collaborative effort in reaching a solution. The petition asks the court to send an investigation team, thus implicitly displaying trust in the team's ability to reach the truth. It highlights the fact that the villagers have faith in the court's ability to discern injustices.

In a good story, says John Truby (2008), the hero and the opponent believe they are right – both have reasons for believing so, and they are both wrong in different ways. One of the killings is of a husband and wife, former Maoists who met in a Maoist camp and fell in love:

> In sworn affidavits, the mother, Tati Bandi and the brother, Tati Ramalu, claim that the deceased couple had been living in Karenar and engaging in farmwork for the past five years. Prior to this, they had indeed joined the banned Maoist party for a period of six months to a year, but having fallen in love with each other, they decided to marry each other and leave the party.

Doubt is introduced in the mind of the listener about whether the deceased couple really left the Maoists. However, admitting to the victims' past links with Maoists creates the impression that the victims had nothing to hide.

By stitching these elements together, the petition exposes the judges to experiences and facts that do not fit the prevailing narrative. For Adivasi, on the other hand, for whom law is both 'a distant but overwhelming power and close at hand to which local desires can be addressed' (Das and Poole 2004), the process of creating the petition familiarizes them with the language of the constitution, and of the dominant nomos.

Transformation

The concrete power of imagery, the use of description, and highlighting the plight of the victims' families whose cries for justice go unheard are the elements that could tip persuasion into transformation. There is, of course, no guarantee of success. All the petition can do is tell a story with a powerful moral problem. Powerful stories, according to Truby (2008), deal with a central moral problem. In *LA Confidential* by James Ellroy (1990), each of the three police detectives has a different answer to the problem of administering justice. One takes the law into his own hands, another makes arrests for money and a third sacrifices justice to get ahead in the political game.

The moral question tackled by the petition is: Can a democratic state seize the fundamental right to life from a citizen without following the procedure of the law? The security forces argue that it can, because the victims were Maoists who shot first, and by killing them in 'self-defence' the forces were protecting the state. The petition says no; because the victims were innocent and had been executed on false pretexts, there has been a misadministration of justice. Furthermore, their fundamental right to life has been taken away by the state without following a procedure established by law.

By slowly exposing the court to multiple accounts that tell similar stories of extra-judicial killings (more individual accounts were added to the original petition, and the case is now being heard with similar cases), the court begins to recognize that more and more of these experiences do not fit the reigning legal paradigm until ultimately, 'the courts are led to reformulate the available models to fit the experiences they had previously ignored' (Winter 1989: 2275).

To achieve such transformation is very difficult for a counter-story. Here, their engagement with the media and civil society groups outside the courtroom can help. For instance, in an interview, Suneeta said that since the petition was filed, the police had hounded them, harassed them and broken up parts of their house. Even as they were helping villagers record their affidavits in the Bijapur court for their court case, local police mounted a door-to-door search for them – forcing them to flee and approach the High Court for interim relief in ensuring their own safety. With the High Court order in hand, these two young women returned to their village, only to find that the police parties had returned twice to their village, broken four homes and beaten up three people.[21] These events were described in a matter-of-fact way, as if it were part of her (and the villagers') day-to-day life to experience such cruelty from the very forces that were there to protect them. The absence of melodrama makes it resonate more powerfully with the listener from the privileged classes (including the judges) because it makes them compare their (secure) reality with this one. It may even evoke outrage.

This process of engaging with the courts, the media and civil society performs two important tasks in achieving transformation. First, it helps the Adivasi to reflect on the collective harm. At a press conference, Suneeta asked:

> How can we be seen as doing something wrong when we are raising the problem of sexual assaults and killings of Adivasi women and men by police and armed forces through approaching the courts and using legal means? Isn't that the best way to seek justice?

These words show how individual harms, when viewed as collective harms, help Adivasi women identify their constitutional right to justice. It brings in the political dimension highlighted by Joanna Wheeler (Chapter 1, this volume).

Second, these interviews outside the courtroom explain the Adivasi ICM to a non-Adivasi public that includes judges. Far from being insulated from politics or society, a judge is susceptible to the fluctuating influences of political machinations, public opinion and national crisis. Where civil liberties and social rights are at stake, India's higher judiciary functions more as an 'embedded negotiator', constantly negotiating with their identities as judges, citizens and members of a state institution (Shankar 2009). As an eminent judge said, 'the court's strength lies in public confidence and support'. After experiencing a loss of this legitimacy due to their capitulation to an emergency regime in 1975–1977, the judges of the apex court brought in several measures such as relaxed standing requirements through PILs and more favourable treatment of non-justiciable social rights. This means that how the story plays out outside the courtroom is just as important in influencing the transformation.

Conclusion

Seeing courtrooms as a place where stories in petitions are designed to appeal in particular ways to the cognitive process helps us see how those at the margins can use these techniques to tell a powerful counter-story. It illuminates how democratic citizenship is not a static event where rights are conferred by a constitution. Rights have to be claimed, and the telling of a counter-story in court and outside of it is one way of doing this. For the Adivasi, engaging with the legal process allows them to move from individual stories of trauma to a collective and political space where an Adivasi understands her own position as a citizen with specific expectations from the Indian state. The trauma becomes representative of a wrong done to a citizen of India. This sentiment is alluded to by Munni Pottam at a press conference organized by the WSS: 'Adivasi people only experience brutality from the state and police. We are approaching the courts hoping we will get something different'.

If counter-stories can quicken and engage the conscience (Delgado 1989), then such stories, when told in democratic settings, have several advantages over authoritarian ones. Democracy allows institutions such as courts to create new modes of reaching the citizens (PILs) and creates strong checks on bad behaviour from other institutions. Second, it allows civil society groups to flourish and forge links with the marginalized. Third, it creates possibilities of new alliances for the marginalized who want to tell their story to the state, as it did for the Bastar villagers: 'The organisation, Women Against Sexual Violence and State Repression, had come to our village as part of a fact-finding team in 2015', said Munni in an interview, narrating how they forged links with the NGO.[22] 'They had come there to assess and record instances of sexual violence against women. We were always raising our voices, just that we didn't know how to take the legal route … We are not alone; if you come to our village, you will realise that no one is scared to speak the truth. They all will say exactly what we are saying'.

As of this writing, the fate of Munni and Suneeta's petition rests with the Supreme Court of India. Though the Supreme Court of India was unable to address the plight of the migrants stranded during the lockdown, it does not mean that the marginalized will always be left out of citizenship rights or that judges will always function in a jurispathic manner. Our analysis of Munni and Sunita's case suggests that regardless of whether or not they are successful in court, their engagement with the legal processes has created pathways for them to articulate their conceptions of justice, their nomos, within the constitutional framework of post-colonial India. While more often than not, legal meaning emerging from judgements favours the non-Adivasi, the analysis presented in this chapter shows us that there are other benefits to be accrued when a marginalized community approaches the court. The legal arena encourages the creation of stories that put individually experienced traumas into a language that law understands and can do something about. It also creates the ability for the marginalized to use the very processes that usually curtail the telling of a normative story to tell a strong counter-story with a strong moral question, and to find overlaps with conceptions of justice that are intelligible to the dominant nomos. Finally, it reveals new pathways to citizenship for the marginalized that require an active and dynamic engagement with the institutions of the state and with other citizens. By engaging with the legal process and creating a powerful counter-story, the Adivasi inhabit a dynamic space of citizenship where the post-colonial democratic state provides the framework but agency rests with the citizen.

Notes

1 I am grateful to Joanna Wheeler, Steven Hartman and Fiona Anciano for their comments and suggestions. I would also like to thank the participants in the workshop on Moral Economies of Participation and Protest, Cape Town, 2018.

2 Milovanovic (1988) describes what gets lost in translation when inmate jailhouse lawyers rewrite what happened in the streets into the language of the courts – loss of race, class and gender-based inequalities that contributed to feelings of anger, alienation and despair in the inmate.

3 According to an expert committee on tribal health, 55% of India's 104 million tribal population live outside the 809 tribal majority blocks. About 45.3% of the tribals in the rural areas live below the poverty line. https://economictimes.indiatimes.com/news/politics-and-nation/reverse-migration-of-peoples-due-to-lockdown-may-destroy-indias-tribal-communities/articleshow/75027171.cms.

4 See Delgado (1989). Also see Lisa Sarmas (1994), who uses *Louth v Diprose* to evaluate the strategy of telling outsiders' stories to achieve progressive social change.

5 Cover highlights an insight of the Talmudic scholar Joseph Caro, that the universalist virtues we have come to identify with modern liberalism, which underpin laws in liberal democracies, are essentially system-maintaining 'weak' forces. They are virtues justified by the need to ensure the coexistence of worlds of strong normative meaning, namely culture-specific designs of particularist meaning (e.g. religious law, traditional codes) (Cover 1983: 16).

6 A note on terminology: 'Indigenous' is problematic because there is a dispute over who is indigenous in India. 'Tribal' has negative connotations, as does the state's term 'scheduled tribes'. Adivasi, forest dwellers, is a more apt way of characterising these groups, and that's the term this chapter uses. See Pooja Parmar (2015) for a discussion on the issues associated with the terminology.

7 *IN THE HIGH COURT OF CHHATTISGARH AT BILASPUR WRIT PETITION (PIL) NO./2016, In Re Extra Judicial Killings in District Bijapur*. All discussions on the 'Petition' refer to this document.

8 Census of India 2011.

9 Naxalites are members of an armed revolutionary group advocating Maoist communism. Their name comes from Naxalbari, a village in West Bengal where an uprising took place in 1967.

10 One such case was *Nandini Sundar v State of Chattisgarh*, where the petitioners were a university professor, a public intellectual and a retired forest official who petitioned the High Court to investigate and stop the government from arming and deploying vigilante groups in the tribal districts. The Supreme Court said that the creation of the Salwa Judum was an abdication by the state of its constitutional responsibility to provide security to citizens and asked the state to desist, stop and disband the Salwa Judum. But has the state done so? No, says Nandini Sundar (2016: 381): 'Police-sponsored vigilante groups … intimidate and drive out journalists, researchers, lawyers and activists'.

11 https://preprod.sabrangindia.in/article/two-young-adivasi-women-hrds-demand-justice-chhattisgarh.

12 http://sanhati.com/excerpted/17774/.

13 Posner (2009) points out that the structure of the judge's opinion has – like a detective story – a double narrative. It goes backward in time, the events that have brought the parties before the court for judgement, and forward in time to tell the story of the court's resolution of the dispute. For sentencing, however, more about character is admissible in some courts. The US Supreme Court allows the defendant to tell a 'no holds barred' story to the jurors to persuade them not to send him to the gas chamber, and the victim's family can also do the same from the victim's point of view. Posner (2009) calls it 'Competing sob stories' and points out a pitfall of such stories – that of atypicality and unrepresentativeness.

14 Research with Aboriginal English speakers has found that silence is an important and positively valued part of many Aboriginal conversations (Eades 2013).

15 As Eades points out, a number of studies have found that there are more Yes/No (or closed) questions in cross-examination (which further complicate and add inconsistencies in the witness's story), and more open-ended questions by the examiner-in-chief. She shows how Australian defence attorneys exploit differences between the usage patterns of Aborigine English to impugn the credibility of young Aborigine speakers. In courtroom situations involving bilingualism, Hale (2006: 156–7) found that interpreters translating Spanish into English regularly altered the form of a witness's answers, often substituting powerless forms (such as hesitations and grammatical errors) that were 'detrimental to the evaluation of the witness's character and credibility'.

16 A similar situation exists for other marginalized peoples globally (Escobar 2011). Retaining a language is part of the imagined future of an Amazonian indigenous person, and the ability to speak other languages ought not to erase that identity. Still, Brazilian courts have stated that if someone can read or speak Portuguese, is able to vote or even knows how to ride a motorcycle, then he is not indigenous and not entitled to any legal benefits derived from this condition (Vitorelli 2014).

17 Vasudha Dhagamwar (2006) argues that an incomprehensible legal system initiated by the British and carried on after independence by the Indian state deprived the Adivasi of their history and identity, and of their land and resources.

18 Delgado (1989) highlights the gap between the way blacks and whites in America think about the inequality between the two – which is illustrative for our case. Many in the white majority hold that the inequality is because of a cultural lag or inadequate enforcement of currently existing beneficial laws, both of which are easily correctible. For many minority persons who are black, the principal instrument of their subordination is neither of these. Rather, it is the prevailing mindset, by means of which the dominant group justify the world as it is, with whites on top and blacks and browns at the bottom. Substitute Adivasi for black, and the non-Adivasi elites for white, and that is the situation prevailing in India.

19 Winter (1989) illustrates the concept of the ICM with this example: Suppose, for example, we enter a restaurant, seat ourselves, and are then confronted by a human with pad and pencil. Does this stranger want to hear our life stories? Challenge our right to enter the premises? Any of these are possible and, on particular occasions, may in fact be the case. Yet, we 'know' automatically that the person is asking for our order because we have unreflexively assumed a 'restaurant scenario', which organizes our understanding of the events around us. Thus, although the scenario is socially contingent, it nevertheless provides sufficient constraints to ground intersubjective meaning because it is grounded in the particular social experiences of members of that culture.

20 Source: https://wssnet.org/2018/01/12/delhi-press-conference-video-of-suneeta-pottam-and-munni-pottam-on-extra-judicial-killings-in-chhattisgarh/.

21 In 2017, nearly a hundred Adivasi villagers in Chhattisgarh tried to file FIRs under the Atrocities Act against what they said was the forced dispossession of their land through threats, intimidation and misinformation by agents acting on behalf of private companies. The Scheduled Caste and Scheduled Tribe police station in Raigarh accepted the complaints at first, but then used the ruse of a 'preliminary inquiry' to delay the investigation. See http://www.khabarichidiya.com/2018/01/10/young-women-spoke-about-these-cases-of-encounters-along-with-the-details-of-the-very-recent-physical-and-sexual-assault-of-the-the-women-of-the-villages-where-suneeta-and-munni-live-press-conference.

22 https://www.nationalheraldindia.com/cafe/activism-comes-at-a-cost. For more interviews, see this link for their press conference in January 2018: https://www.youtube.com/watch?v=CtmDDqqe7u4.

References

Census of India. 2011. Available at https://censusindia.gov.in.

Cooke, M.C. 2002. 'Indigenous Interpreting Issues for Courts', *The Secretariat – Australasian Institute of Judicial Administration*. Retrieved 10 June 2019. http://www.naclc.org.au/cb_pages/files/Cooke%20-%20Indigenous%20interpreting%20issues%20for%20courts.pdf.

Cover, R. 1983. 'Foreword: Nomos and Narrative', *Harvard Law Review* 97(4): 1983–1984.

Dannenberg, H.P. 2008. *Coincidence and Counterfactuality: Plotting Time and Space in Narrative Fiction*. Lincoln: University of Nebraska Press.

Das, V., and D. Poole. 2004. *Anthropology in the Margins of the State*. New Delhi: Oxford University Press.

Delgado, R. 1989. 'Storytelling for Oppositionists and Others: A Plea for Narrative', *Michigan Law Review* 87(8): 2411–2441.

Dhagamwar, V. 2006. *Role and Image of Law in India: The Tribal Experience*. New Delhi: SAGE Publications.

Eades, D. 2013. 'Telling and Retelling your Story in Court: Questions, Assumptions, and Intercultural Implications', in D. Eades (ed.), *Aboriginal Ways of Using English*. Australia: Aboriginal Studies Press, pp. 162–187.

Ellroy, J. 1990. *L.A. Confidential*. New York: Grand Central Publishing.

Escobar, A. 2011. *Encountering Development: The Making and Unmaking of the Third World*. Princeton: Princeton University Press.

Hale, S.B. 2006. *The Discourse of Court Interpreting*. Amsterdam: John Benjamins Publishing Company.

Milovanovic, D. 1988. 'Jailhouse Lawyers and Jailhouse Lawyering', *International Journal of the Sociology of Law* 16(4): 455–475.

Parkin, A.J., J. Ward, C. Bindschaedler, E.J. Squires and G. Powell. 1999. Recognition Following Frontal Lobe Damage: The Role of Encoding Factors, *Cognitive Neuropsychology* 16: 243–265.

Parmar, P. 2015. *Indigeneity and Legal Pluralism in India: Claims, Histories, Meanings*. Cambridge: Cambridge University Press.

Posner, R. 2009. *Law and Literature*. Massachusetts: Harvard University Press.

Sarmas, L. 1994. 'Storytelling and the Law: A Case Study of *Louth v Diprose* [1994]', *Melbourne University Law Review* 19(3): 701–728.

Scheppele, K.L. 1989. 'Foreword: Telling Stories', *Michigan Law Review* 87(8): 2073–2098.

Shankar, S. 2009. *Scaling Justice: India's Supreme Court, Anti-Terror Laws and Social Rights*. New Delhi: Oxford University Press.

Sundar, N. 2016. *The Burning Forest: India's War in Bastar*. New Delhi: Juggernaut Publications.

Truby, J. 2008. *The Anatomy of a Story*. London: Farrar, Strauss & Giroux.

Vitorelli, E. 2014. 'Linguistic Minorities in Court: The Exclusion of Indigenous Peoples in Brazil', *Language and Law/Linguagem e Direito* 1(1): 159–173.

Winter, S.L. 1989. 'The Cognitive Dimension of the Agon Between Legal Power and Narrative Meaning', *Michigan Law Review* 87: 2225–2279.

9 Including the excluded

Interests and values in the Brazilian public health care system[1]

Vera Schattan P. Coelho

Introduction

From the 1930s to the 1980s, Brazil presented a slow but permanent expansion of its social policies, gradually encompassing a broader section of the population. With the Constitution of 1988, this movement went deeper, and social citizens' rights, such as health and education, became universal. Thus, the 1988 Brazilian Constitution, which established the formal transition to democracy, assured an important set of social rights for all. In doing so, it formalized a new social contract between citizens and the state, in which the legitimacy of the latter was linked to its capacity to assure the development of citizens' capabilities, contributing to providing them with more life opportunities. This included the right to health and, from the 1990s forward, Brazil's universal public health care system, referred to as the *Sistema Único de Saúde* (SUS), was implemented. During the period in which the SUS was expanded, primary health care grew considerably, and a reduction in health inequalities was reported (Hone et al. 2017; Coelho and Dias 2018). In 2013, there were protests throughout Brazil calling for less corruption and better social services. In the following years, a strong political divide became paramount between those in favour of and those opposed to the *Partido dos Trabalhadores* (PT), the Workers' Party and the ruling party at the time. In 2017, after twelve years in power, the PT[2] lost the national elections, and a set of progressive social policies, including health care programmes, were threatened with extinction.

This chapter investigates this trajectory, starting from a perspective closely linked to political science, which understands the expansion of universal social policies as being, on the one hand, associated with the state's capacities and, on the other hand, with coalitions that bring the poor and the middle class on board (Mares and Carnes 2009; Garay 2014). The following questions then emerge: Why did popular movements take to the streets to protest against these policies? And what expectations are there for maintaining these policies in a more conservative setting? To deal with these matters, attention is called to the need to deepen understandings of how the right to health care and the responsibility of the state to provide it have evolved and have been experienced by the citizenry at different moments in history.

To do so, this chapter focuses on how interests, political values and institutions have come together in two moments. During the first moment, forged during the 1970s and 1980s, there was a broad social movement that was fighting against the military regime. The second, more recent moment, was experienced during a democratic regime when less cohesion was observed among the poor, as well as between the poor and the middle class. The discussion highlights the interests and values-based perspectives mobilized around the public health care system during these periods and the contribution, as well as the trade-offs, of different strategies of political participation in guaranteeing access to health services. In this sense, the 1988 Constitution expresses an agreement between different groups concerning the need, in an unequal country such as Brazil, to assure access to health care services as a means to support the development of individual's capabilities and, in so doing, also help to mitigate inequalities. In more recent times, this agreement has come under threat, leading to popular protests demanding the maintenance of social and economic policies.

Interests, political values and health policy

The SUS is widely recognized as a successful example of policy formation and implementation (Victora et al. 2011). Despite this consensus, we still do not have an assertive answer to the following question: Did historically disadvantaged groups gain access to both health services *and* the policy process that defines the decisions made and resources allocated? In fact, the answer to this question has varied over the past thirty years. At times when it seemed that we could answer it with a resounding 'yes', we realized that the challenge of inclusion was still there, re-attached to old and new situations and groups. In the words of Holston (2007: 4), 'the challenge of inclusion and its contestation' remained.

It is this dynamic of change and continuity that is explored in this chapter, with two main objectives. First, by drawing attention to how, in the case of health, the building of a new realm of participation, rights and citizenship has taken place in Brazil, as well as the fact that this realm is currently under threat. This effort is guided by an analytical approach that highlights the importance of political coalitions and government capacities in driving decisions concerning social policies (Esping-Andersen 1991; Mares and Carnes 2009; Garay 2014). Second, exploring how, during this process, the right to health was built as a relational concept – one that establishes a vertical link between citizens and the state – and how this concept drives both state decisions about how, when and who was included in the SUS, as well as citizens' perceptions of the fairness of these decisions. In doing so, the chapter explores the mechanisms that articulated citizens' political values and the state's capacity, which in turn led to defined rights and obligations in the process of the formation and implementation of the SUS.[3]

The narrative presented here is based on a review of the existing literature; different case studies focusing on health policies and programmes at the municipal, state and federal levels; and quantitative assessments, which were coordinated by the author from 2001 to 2016. The historical background that covers the period from 1970 to 1990 relies on a review of literature. Official documents, as well as academic and press articles were collected and interviews were conducted to reconstitute the implementation of SUS's primary health policies. A qualitative case study supported an analysis of the challenges of inclusion experienced by the SUS in Cidade Tiradentes. This case study included participant observation at health facilities and in local communities, as well as interviews during which São Paulo Municipal Health Secretariat managers, health service providers, politicians and citizens responded to closed and open-ended questions. In total, 184 interviews were conducted with health councillors, public officials, health professionals and SUS users at the municipal, state and federal levels.

The chapter begins by presenting a brief overview on the process of building and implementing the SUS (1980s–2010s), highlighting the emergence of new discourses, spaces and networks that allowed different marginalized groups – such as poor urban residents, female, indigenous, Afro-Brazilian and *quilombola*[4] populations – to demand and gain inclusion in both the health policy process and access to health services. In the second section, it is argued that these results, while positive, do not apply, for example, to groups that having recently arrived in poor and violent neighbourhoods on the outskirts of metropolitan areas, live in illegal areas and have very limited access to the SUS. In the third section, the dilemmas faced by the SUS in dealing with populations that are at the margins and the role of health inequalities in encouraging the 2013 protests are explored. Finally, the chapter concludes with a brief discussion of SUS's prospects. It is expected that the analysis, based on both an understanding of the political values that supported the social contract agreed upon in 1988, and the preferences of the ruling coalition and state capacities, will help to predict the level of resilience of the SUS within the present political context.

The building of the SUS, a public universal health system

To understand the policy process that made it possible for thousands of citizens and interest groups to be able to work throughout the country gathering support to demand the creation of the Brazilian public health care system and its participatory spaces, we need to go back to the 1960s. From 1964 to 1984, Brazil lived under a military regime that restricted political rights. However, an expansion of social rights was seen in this period. According to Dowbor (2008), despite the dominance of a hospital-centric model during this period, there was room for actions at the level of the Health Ministry which enabled alternative interests to flourish. Its meagre resources left this Ministry on the sidelines and thus more open to the trends of community medicine and innovations in governance. These trends were aligned with the approach of international health organizations and were an alternative to the emphasis on hospital-based medical care. In this context, public health physicians, who occupied management positions within the Health Ministry, had relative freedom in staffing their teams and initiating coverage-increasing programmes, especially in rural areas (Dowbor 2008).

During this time, the sanitary movement became a collective actor, encompassing several health care professionals – union-organized doctors, academics and the medical student movement (Barros 2002; Escorel et al. 2005; Dowbor

2008; Coelho and Waisbich 2016). They occupied posts in public health institutions, advocating for the effective universalization of the health system and the institutionalization of citizens' participation in the formulation, management and monitoring of health policy (Neder 2001). The sanitary movement encouraged the mobilization of a larger health movement. The later movement's trajectory had a lot in common with other popular movements that started during the period of Brazil's re-democratization. It was organized around Catholic church pastors, neighbourhood societies, the Basic Christian Communities (*Comunidades Eclesiais de Base* in Portuguese), mothers' clubs, popular education groups and movements against poverty (Singer and Brant 1983; Sacardo and Castro 2002). The PT, which was a new political actor in the early 1980s, also collaborated with the health movement by providing places for meetings, opening up institutional space for popular movements and supporting candidates connected to communities and the health movement (Machado 1995; Bógus 1998). It was the collaboration of the sanitary movement with intellectuals, students and artists that played an important role in 'creating, with their work and social influence, the political conditions and infrastructure for popular causes', and so formed 'a support network for the health movement' (Neder 2001: 94).

For example, in 1977, a group of women from different poor neighbourhoods in São Paulo's eastern zone began to organize public events and protests to pressure the state Health Department into constructing a health centre in the region (Neder 2001). Looking at these events, Cardoso (1982) recognizes similar mechanisms to the ones described earlier. According to her, since 1974, local leaders together with the Basic Christian Communities and other political and professional groups helped to organize meetings and debates in which the notion of being a 'community' was developed in the region. This notion, which was developed through the sharing of a common experience – the condition of being poor and living under difficult circumstances in the periphery of the city – was key to concealing conflicts within the organization of the popular classes. In this environment, the idea of equality among citizens supported growing claims for social justice and the guarantee of equal access to health care. The health centre that was demanded was inaugurated in 1979, and a health council for the centre was created that same year. Members of the council, who would act as a monitoring group, were elected by the 8146 citizens living in the region (Bógus 1998). This led to another eighteen health councils being elected to monitor health centres in the eastern zone in 1981.

These movements continued to grow during the democratic transition of the government that succeeded the military regime in 1985. During this phase, representatives from the sanitary movement took on new key positions at the institutions in charge of health policy in the country, reinforcing the principles of change: decentralization and, more specifically, municipalization of health services; universal and equal access to health services; comprehensive care; regionalization and integration of services; and the development of collegiate institutions (Noronha and Levcovitz 1994; Lima et al. 2005; Dowbor 2008). Equally noteworthy was the role played by civil society organizations. These organizations were engaged in a number of local initiatives, such as the creation of HIV/AIDS programmes, and also helped to disseminate the notion of health as a citizen's right (Nunn 2009; Coelho and Lavalle 2019).

In 1988, when the Constitution was enacted and the right to health became a universal right, the Brazilian health sector had already travelled the long road of reforms that universalized free access to health care and decentralized services to states and municipalities, albeit while maintaining the private sector as one of the service providers (Dowbor 2008).[5] It is important to keep in mind how the relational concept of the right to health evolved during this period. Authorities, health professionals and citizens built a shared understanding of what was fair in terms of the right to health. This was unlike what had happened previously, when access to health services was dependent on financial contributions. Representing a shared set of political values, health became understood as a right of all citizens and a duty of the state, which was expected to guarantee free access to quality services for health promotion and recovery.

This brief retrospective points to a process through which the poor were able to make political alliances that assured them greater access to public health services, as well as to the policy process that, between 1980 and 2000, helped to create spaces and networks that organised different social segments around the health movement and its insistence on health as a citizen right.

Despite these significant advances, the challenges posed to the inclusion of historically disadvantaged groups had become clearer by the mid-1990s, with increased demands for health policies to meet the specific needs of indigenous, black and *quilombola* populations, who had the worst health indicators in the country. For example, in the 1990s, the likelihood of a black child dying before completing one year of life was 1.8 times higher than for a white child, while maternal mortality among black mothers was 3.6 times higher than for white mothers (Cunha 2008). Menegolla (2008) reported that infant mortality among indigenous children was double the rate for white children, while tuberculosis rates among indigenous people were three times higher than in the overall population.

During the 1990s (with the guarantee of health as a citizen right, as stipulated in the 1988 Constitution), there was a gradual change in the health paradigm to focus on integral primary health care, a concern with equity in health care and the growing recognition of diversity. This focus not only drew attention to the specific characteristics of women, black, indigenous and *quilombola* health issues; it also resulted in an increase in the number of programmes and projects for these populations. For example, in 1992, the first 'Special Indigenous Health District' (DSEI) was created in an area inhabited by the Yanomami people of the Northern Amazon in an effort to halt high mortality rates due to malaria and other infectious diseases that were occurring as a result of invasions by wildcat gold miners. In 1999, the Indigenous Health Subsystem was created based on the special indigenous health districts – units designed to reflect the fact that indigenous territories crossed municipal and sometimes even state boundaries. In total, thirty-four special districts were created, each overseen by a council, of which half of the members were to be representatives of indigenous health service users as stipulated by law. Parallel movements were formed in the 1990s and 2000s by the feminist movement, Afro-Brazilians and *quilombolas* (Coelho and Lavalle 2019).

In all cases, these efforts contributed to the enactment, during the 1990s and 2000s, of policies that recognize the specificities of these groups and seek to respond more effectively to their health needs. To understand what made this achievement possible, we need to keep in mind that the indigenous and black movements had been able to establish complex relations with political parties; links with different global, national and regional organizations, such as Oxfam, Gaia Foundation, Health Unlimited, Indigenist Missionary Council (CIMI) and others; as well as with different sections of the government at the federal, regional and municipal levels. Through their involvement in regional and national meetings and also in more institutionalized forums, councils and conferences, these movements and their organizations had succeeded (at least partially) in participating in the health policy process. Once these channels were in place, they helped to build alternative perspectives and to create bridges between different cultural practices (Coelho et al. 2014; Coelho and Lavalle 2019).

These mobilizations and progressive policies and programmes happened in tandem with a growing emphasis on primary care, the implementation of the Family Health Programme and other programmes that prioritized poor regions and population groups. As a result, immunization coverage had increased significantly, from around 50% in 1980 to nearly 100% in the early 2000s (Coelho and Dias 2018). Between 1996 and 2006, the proportion of pregnant women who had no prenatal doctor visits dropped from 26% to 1.3% (Ministry of Health 2010; Gragnolati et al. 2013). These initiatives – along with access to safe water, sanitation and quality food – in addition to the economic situation of households, ensured that when it came to inequalities, the poorer north-eastern and northern regions experienced the largest reductions in infant mortality, although they continue to have higher infant mortality rates.[6] Disparities in health outcomes have been reduced not only geographically, but also across socioeconomic groups. For instance, the reduction in the infant mortality rate was much greater among low-income groups, contributing to the convergence of different income groups to around 20 deaths out of 1000 live births by the mid-2000s (Gragnolati et al. 2013; Coelho and Dias 2015).

This promising picture explains an important part of what has happened in Brazil's health sector in the last thirty years. There is a universal system anchored in the Brazilian Constitution, which has prioritized the poorest, invested heavily in primary care and achieved promising results. From an analytical perspective, we have seen: (1) a cohering set of political values with regards to health care, resulting in an important alliance between the poor – especially the urban poor and the middle class – in favour of public health care expansion; (2) this expansion aimed to guarantee more equitable access to health promotion and recovery, helping to tackle deeply entrenched health inequalities between racial, ethnic, gender and socio-economic groups; (3) a growing recognition – shared by political leaders, health authorities and citizens – of health as a right, helping to sustain the development of universal health policies aimed at ameliorating the health conditions of low-income groups; and (4) the state's capacity to successfully implement health programmes and improve health indicators.

It would be hypocritical, however, to conclude this chapter without mentioning another reality taking place in parallel, which Holston (2007) describes well, concerning the new forms of violence and exclusion that simultaneously erode the notion of health as a right. Exclusion and violence are incredibly resilient. To understand this reality, I highlight and discuss the situation of Cidade Tiradentes, a poor peripheral area in the city of São Paulo and one of the centres of both the sanitary movement and the health movement.

The permanent challenge of inclusion

Fifty years ago, the area of Cidade Tiradentes was occupied by Atlantic Forest, and in the 1980s, over 40,000 homes were built there (the biggest public housing complex in Latin America) in order to house industrial workers (Rede Nossa São Paulo 2015). The workers that moved into these buildings found a neighbourhood that lacked sanitation, public transportation, schools, health services and so forth. To address this situation, they began to organize into different associations. Two important associations were the Neighbourhood Association and the Church. They organized regular meetings to discuss the situation and to find ways to demand public services and to pressure the state and municipal authorities. For example, petitions were written and *caravanas* (retinue/entourage) were organized to deliver the documents to the municipal government. Sometimes these *caravanas* would initiate conversations with authorities, who ended up visiting the area and promising to provide hospitals, schools, public phones and other public goods. These mobilizations helped to call attention to the area and, once democracy was in place, a number of political candidates and authorities worked to consolidate the area's infrastructure.[7]

Today, 220,000 inhabitants live in the area, and the black community makes up over 45% of the population. Its disorganized occupation has created two levels of poverty: one formed by formal workers, comprised of 160,000 people living mainly in the buildings that form the public housing complex, and where the head of household earns an average income of $150 to $350 per month; and an informal group who illegally occupy protected areas of the remaining Atlantic Forest, with household members earning from $30 to $150 monthly, and where the illiteracy rate is widespread at 20% of the 60,000 inhabitants, as well as high rates of teen pregnancy (Rede Nossa São Paulo 2015).

Figure 9.1 shows the neighbourhood of Cidade Tiradentes, one of the 31 sub-municipalities of the city of São Paulo, where 12 million inhabitants live. Today, Cidade Tiradentes has good health infrastructure, ten basic health units, one hospital, two outpatient clinics and one emergency room. Also importantly, health councils are organized and active in health units and in technical supervision.[8] These councils meet monthly and regularly monitor health services and health policy. This infrastructure has been expanded considerably over the last twenty years and has contributed to the reduction of inequalities in access to health services between the city's better-off and worse-off areas (Coelho et al. 2019).

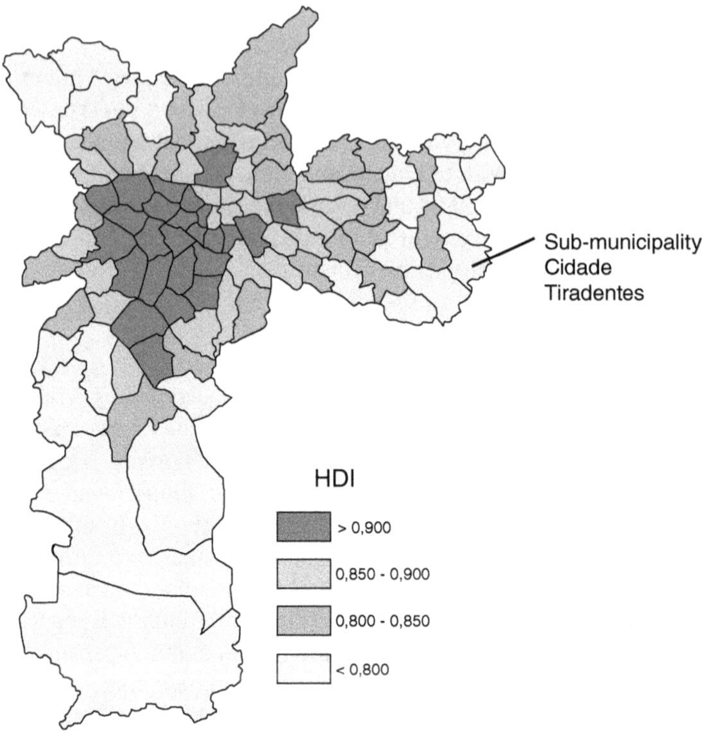

Figure 9.1 Municipality of Sao Paulo, indicating the Human Development Index by sub-municipalities, highlighting the sub-municipality of Cidade Tiradentes, 2019.

The sad news is that this structure is not serving the extremely poor people who live in the illegal areas of Cidade Tiradentes. They remain on the border between neighbourhoods – for example, an area of environmental preservation that is illegally occupied – and none of the health authorities have taken responsibility for them. The inequality and exclusion they experience is persistent. The inclusionary infrastructure that supported – and still supports – the project of building a universal SUS capable of tackling inequalities has not been able to include the 'new' excluded, a group largely made up of foreign immigrants and extremely poor Brazilians.

Revisiting the challenge of inclusion

The stories presented in the previous sections point to important victories. Civil society members (the poor, marginalized citizens, intellectuals, students, etc.), professional societies (doctors, lawyers, anthropologists, etc.), politicians and policy makers formed networks and coalitions to promote a progressive agenda of health for all that contributed decisively to the enactment of a universal public health care system in 1988. In subsequent years, it also contributed to the deepening of the ability of this system to serve poor and marginalized citizens. These achievements happened in tandem with the institutionalization of the Brazilian health movement, which meant that a group of its leaders (who defended a universal and decentralized health care system), assumed powerful positions inside the government while another group of them, linked to community leaders who mobilized for health services during the 1970s and 1980s, moved to the network of health councils that have been organized all over the country.

These stories show a process where, between the late 1970s and 2010s, coalitions of social actors fighting for democracy and the reduction of inequalities, as well as different projects – health reform, self-recognition of indigenous and black populations, women's rights, *quilombolas* and LGBT issues among many others – have played a central role in the establishment of political and social rights and the attainment of policies to meet their needs. They also point to the central role played by health authorities and state institutions in defining and implementing policies to respond to these claims. Nevertheless, the Cidade Tiradentes case reported here shows that these successful stories co-exist with the perpetuation of exclusion and inequality. Although huge gains were made for the poor population living in legalized areas of the region, not even the SUS has been able to treat those living in some of the country's illegal areas with dignity.

While this case study has presented a successful story of building a universal health system that has helped to challenge inequality, it has also made it clear that the included-excluded dichotomy remains present, even in the poor peripheries which, like Cidade Tiradentes, have gained access to a well-structured primary health system over the last twenty years. This situation is even worse in other settings, such as rural areas that are isolated and offer a very limited range of services. To understand these exclusions, it is important to bear in mind that the public health system faces a number of operational, legal and cultural constraints in organizing services in remote, illegal or culturally diverse areas. Nevertheless, while these may be recognized as fair reasons to justify gaps in service delivery, they may not work to stop rising popular indignation due to what is perceived as an unfair distribution of services. The perception of unfairness comes from the fact that some groups are recognized as being privileged over others that are more in need. This indignation is probably also linked to the perception of the low quality of services offered by the SUS, which is worse in more peripheral areas, vis-à-vis those offered in more central areas, as well as those offered by private companies to high-income groups. The imbalance in the distribution and the quality of public health services, and the perception that some groups have been privileged over others, may have been important fuel for the 2013 protests. The indignation, linked to the perception that the agreement to ensure access to quality universal health care services remains unfulfilled, may help to explain why, during a period of SUS expansion, greater access to health services and the policy process as well as inequality reduction, citizens took to the streets to demand better services.

In the following years, important changes have taken place in the Brazilian political scenario. In 2015, Dilma Roussef was re-elected in a hotly disputed election.

In 2016, she was impeached and Michel Temer, the Vice President, assumed the presidency. During his term, within the context of severe economic crisis, fiscal austerity became the norm in the realm of social policies. In 2018, Jair Bolsonaro, a right-wing politician with weak links to the traditional party structure, was elected president. The profile of his voters is described as being more white, more educated and wealthier than that of PT voters (Folha de São Paulo 2018). His electoral promises include a revision of a number of social policies focused on minorities and marginalized social groups.

These radical changes in the political scenario pose doubts where the future of the SUS is concerned. According to Mares and Carnes (2009), in countries with government capacities similar to the ones found in Brazil, the shift in the ruling coalition from a coalition of the middle classes and the poor to one of middle and upper classes will mean that contributory policies will gain support to the detriment of the current universalistic policies. From my own perspective, the questions that remain are: Is a significant part of Brazilian society wedded to the cause of health as a right and a duty of the state? What will their reaction be to such a programmatic change? How will they react to the breaking of constitutional agreements?

Final remarks

The process of the creation and democratization of the SUS may be described as the result of a process that demanded cohering political values around health care, political agency and social energy. Agency and energy were present during the re-democratization process, when a coalition of people from different social and professional backgrounds came together to fight against the dictatorship. At that time, the progressive Catholic church was stronger, the industrial working class had a stronger economic role and intellectuals had been intimidated for many years. This coalition was also present during the first years of democracy and prompted a sense of urgency in fighting dictatorship and inequality. The social contract that emerged during this period played a central role in generating policies and practices that have contributed to creating a universal health system that is also accountable to poor and marginalized populations. In addition, the favourable financial conditions experienced during the 2000s and state capacities present at the various levels of the federation facilitated the implementation of this project.

Despite important achievements in reducing health inequalities and in guaranteeing greater access to public health care services, there is a resurgent popular feeling, expressed on the streets in 2013, that the system is not delivering accordingly. In the following years, a new coalition has made gains in the national elections. In the realm of social policies, major revisions have been promised. One of the first initiatives of the Bolsonaro government was to slash resources allocated to educational programmes. Immediately, there was a strong popular reaction, with people protesting on the streets throughout Brazil. This can be seen as a good thermometer of what may happen if similar actions are taken in the realm of health policies.

In short, SUS's trajectory underscores the strength of a social contract that establishes relations between rulers and the ruled, who support the legitimacy of the former. There is a shared idea of what a fair distribution of services is and what rulers' responsibilities are when it comes to providing health services. This is a central dimension to understanding not only why and when citizens go to the streets to protest, but also the direction of policy changes.

In this chapter, it is argued that to take into account political values, demands for inclusion and social justice, and the alignment of civil society interests may lead to predictions about SUS's future which differ from those stated by authors focused on political coalitions and state capacities. While the latter group of authors predict a move to contributory health policies, the former suggests that radical changes to the social contract that has historically given the SUS its legitimacy will not happen in the near future, as challenging its universalism would probably prompt strong mass mobilization and resistance. A more balanced approach, which looks at coalitions' interests, citizen's values and state capacities, including financial availability, suggests that the changes that are likely to happen are managerial moves towards more privatized modes of service delivery, which may partially answer the demands of the middle and upper classes for less state intervention and more efficiency, without challenging the right to health paradigm.

Notes

1 This chapter presents the results of the following projects: "Equity and Contracting out SUS's Services" supported by Fapesp process 2013/07616-7 and coordinated by the Centre of Metropolitan Studies; "SUS's Multilevel Governance" supported by CAPES, a post-doctoral grant to Vera S.P. Coelho at the Federal University of ABC; "Learning from International Experiences" supported by TARSC/RWJ foundation and "Accountability for Health Equity in Brazil and Mozambique", supported by ESRC/DFID and coordinated by the Institute of Development Studies, University of Sussex. The results hereby presented were developed in partnership with the Citizenship, Health and Development Nucleus of the Brazilian Centre of Analysis and Planning (CEBRAP).

2 PT was elected for four consecutive tenures, which amounts to a total of 16 years. In 2016, during the last tenure, Dilma Roussef, the President, was ousted following an impeachment process. Michel Temer, the Vice President, who was affiliated with another party, assumed the presidency.

3 This chapter does not contain a discussion of the importance of path dependence in facilitating or constraining political decisions, which is certainly a very important variable in policy development. For a more accurate discussion on its importance, see Paul Pierson (2000), Theda Skocpol (1992) and Hugo Heclo (1974).

4 *Quilombola* is a common name for descendants of black slaves whose ancestors, during the period of slavery, fled from the sugarcane mills, farms and smallholdings on which they performed various menial jobs to form small refugee villages called *quilombos.*

5 Even today, there is an important dichotomy between the public and private sectors in Brazil. The health sector represents 6.6% of Brazil's Gross Domestic Product (GDP). The public sector takes care of 70% of the Brazilian population and spends 47% of this 6.6%, while the private sector covers 30% of the population and expends 53% of the GDP (Coelho 2015).

6 Between 1980 and 2010, the standard deviation among states' infant mortality rates fell by 84% (from 24.6 to 3.8). Meanwhile, in terms of life expectancy, the drop was

33% (from 3.3 to 2.2 years). This narrowing in inequalities can also be seen in the coefficient of variation, which in the same period fell by 33% for infant mortality and 40% for life expectancy at birth. These indicators also improved strongly among the municipalities (Coelho and Dias 2015).

7 Interviews were provided to the project by long-time residents of the area who played active roles in social mobilization from the 1970s to the 1990s, and who more recently served as health councillors on the local health councils.

8 In this sense, it is important to keep in mind that in the city of São Paulo there is a municipal health council, and in each sub-municipality there is a local health council. Inside the sub-municipality, within each health unit, there is a unit health council (Coelho 2015).

References

Barros, M.E. 2002. *Financiamento do Sistema de Saúde no Brasil: Marco Legal e Comportamento do Gasto, Projeto de Desenvolvimento de Sistemas e Serviços de Saúde*. Brasília: Pan-American Health Organization.

Bógus, C. 1998. *Participação Popular em Saúde*. São Paulo: Annablume.

Cardoso, R. 1982. 'Comunidade e Movimentos Sociais Urbanos', in T.P.R. Caldeira (ed.), *Obra Completa*. Rio de Janeiro: Mameluco.

Coelho, V.S. 2015. 'Local Government', in G. Mazzoleni, K.G. Barnhurst, K. Ikeda, R.C.M. Maia and H. Wessler (eds.), *The International Encyclopaedia of Political Communication* (Vol. 3). New Jersey: Wiley-Blackwell, pp. 1–6.

Coelho, V.S., and M. Dias. 2015. 'Saúde e Desigualdade no Brasil', in M. Arretche (ed.), *Trajetória das Desigualdades: Como o Brasil Mudou nos Últimos Cinquenta Anos*. São Paulo: Editora Unesp/CEM.

Coelho, V.S. 2018. 'Health and Inequalities', in M. Arretche (ed.), *Path of Inequality in Brazil*. New York: Springer, pp. 183–205.

Coelho V.S., and A. Lavalle. 2019. 'Os Movimentos Negro e Indígena e a Política de Saúde e de HIV/Aids: Institucionalização e Domínio de Agência' [The Black and Indigenous Movements and the HIV/AIDS Health Policy in Brazil], in A. Lavalle, E. Carlos, M. Dowbor and J. Szwako (eds.), *Movimentos Sociais e Institucionalização: Sociais, Raça e Gênero no Brasil Pós-transição*. Rio de Janeiro: Ed Uerj, pp. 331–373.

Coelho, V.S., L.M. Marcondes and M. Barbosa. 2019. 'Accountability e Redução das Desigualdades em Saúde: A Experiência de São Paulo'. *Novos Estudos CEBRAP* 38(2): 323–349. http://dx.doi.org/10.25091/s01013300201900020003

Coelho, V.S., F. Szabzon and M. Dias. 2014. 'Política Municipal e Acesso a Serviços de Saúde: São Paulo 2001–2012, quando as Periferias Ganharam mais que o Centro', *Revista Novos Estudos CEBRAP* 100: 139–161.

Coelho, V.S., and L. Waisbich. 2016. 'Participatory Mechanisms and Inequality Reduction: Searching for Plausible Relations', *Journal of Public Deliberation* 12(2): 1–16. Retrieved 20 April 2019. https://www.publicdeliberation.net/jpd/vol12/iss2/art13/.

Cunha, E.M.G.P. 2008. 'O Recorte Racial no Estudo das Desigualdades em Saúde', *São Paulo em Perspectiva* 22(1): 79–91.

Dowbor, M. 2008. 'Origins of Successful Health Care Reform: Public Health Professionals and Institutional Opportunities in Brazil', *IDS Bulletin* 38(6): 74–81.

Escorel, S., D.R. Nascimento and F.C. Edler. 2005. 'As Origens da Reforma Sanitária', in Lima, N.T., et al. (eds.), *Saúde e Democracia: História e Perspectivas do SUS*. Rio de Janeiro: Fiocruz, pp. 59–81.

Esping-Andersen, G. 1991. 'As três Economias Políticas do Welfare State', *Lua Nova: Revista de Cultura e Política* 24: 85–116.

Folha de São Paulo. 2018. Pesquisa DataFolha. Retrieved 23 July 2020. https://g1.globo.com/politica/eleicoes/2018/eleicao-em-numeros/noticia/2018/10/03/pesquisa-datafolha-veja-perfil-dos-eleitores-de-cada-candidato-a-presidente-por-sexo-idade-escolaridade-renda-e-regiao.ghtml.

Garay, C. 2014. 'Including Outsiders: Social Policy Expansion in Latin America', PhD Dissertation. Berkeley, CA: University of California.

Gragnolati, M., M. Lindelow and B. Couttolenc. 2013. *Twenty Years of Health System Reform in Brazil: An Assessment of the Sistema Único de Saúde*. Washington, DC: The World Bank.

Heclo, H. 1974. *Modern Social Politics in Britain and Sweden*. New Haven: Yale University Press.

Holston, J. 2007. *Insurgent Citizenship*. Princeton: Princeton University Press.

Hone, T., D. Rasella, M.L. Barreto, A. Majeed and C. Millett. 2017. 'Association between Expansion of Primary Healthcare and Racial Inequalities in Mortality Amenable to Primary Care in Brazil: A National Longitudinal Analysis', *PLoS Medicine* 14(5).

Lima, N.T., S. Gerschman, F.C. Edler and J.M. Suarez. 2005. *Saúde e Democracia: História e Perspectivas do SUS*. Rio de Janeiro: Fiocruz.

Machado, M.H. 1995. *Profissões de Saúde: Uma Abordagem Sociológica*. Rio de Janeiro: Fiocruz.

Mares, I., and M.E. Carnes. 2009. 'Social Policy in Developing Countries', *Annual Review of Political Science* 12: 93–113.

Menegolla, I.A. 2008. *Modelos para o Sistema de Saúde Índigena*. Brasília: FUNASA and World Bank.

Ministry of Health. 2010. *Saúde Brasil 2009: Uma Análise da Situação de Saúde e da Agenda Nacional e Internacional de Prioridades em Saúde*. Brasilia: Ministry of Health.

Neder, C.A. 2001. 'Participação e Gestão Pública: A Experiência dos Movimentos Populares de Saúde no Município de São Paulo', PhD Dissertation. Campinas: Universidade Estadual de Campinas.

Noronha, J.C., and E. Levcovitz. 1994. 'AIS-SUDS-SUS: Os Caminhos do Direito à Saúde', in R. Guimarães and R.A.W. Tavares (eds.), *Saúde e Sociedade no Brasil: Anos 80*. Rio de Janeiro: Relume Dumará, pp. 73–111.

Nunn, A. 2009. *The Politics and History of AIDS Treatment in Brazil*. New York: Springer.

Pierson, P. 2000. 'Increasing Returns, Path Dependence and the Study of Politics', *American Political Science Review* 94: 251–267.

Rede Nossa São Paulo. 2015. Mapa da Desigualdade em São Paulo. Retrieved 21 July 2020. https://www.nossasaopaulo.org.br/portal/arquivos/Quadro_da_Desigualdade_em_SP.pdf.

Sacardo, G.A., and I.E.N. Castro. 2002. *Conselhos Municipais de Saúde*. São Paulo: Pontifícia Universidade Católica de São Paulo, Instituto Polis.

Singer, P., and V.C. Brant. 1983. *São Paulo: O Povo em Movimento*. São Paulo: Petrópolis Vozes.

Skocpol, T. 1992. *Protecting Soldiers and Mothers: The Political Origins of Social Policy in the United States*. Cambridge: Belknap Press.

Victora, C.G., M.L. Barreto, L.M. Carmo, C.A. Monteiro, M.I. Schmidt, J. Paim, F.I. Bastos, C. Almeida, L. Bahia, C. Travassos, M. Reichenheim and F.C. Barros. 2011. 'Health Conditions and Health-Policy Innovations in Brazil: The Way Forward', *The Lancet* 377(9780): 2042–2053.

10 Negotiating foreign policy from below

Voice, participation and protest

Laura Trajber Waisbich

Introduction

Johannesburg, April 2018. A hundred people gathered in the financial district of Sandton to protest against the BRICS grouping (Brazil, Russia, India, China and South Africa) and its New Development Bank (NDB). A year later in Cape Town, a similar crowd took to the streets again, this time at the margins of the fourth NDB annual meeting to protest against the bank – raising environmental, transparency, labour and anti-corruption concerns.[1] Compared to previous BRICS-related activist gatherings, the Johannesburg and Cape Town street protests were relatively small. Street protests brought together only a fraction of the social activists that have been flocking to the so-called 'BRICS counter-summits' year after year since 2013, at the margins of the official summits taking place each year in one of the BRICS member countries (Waisbich 2017). Yet citizens and civil society groups' taking to the streets to discuss and challenge the workings of the BRICS Bank certainly reveals the unfolding dynamics of domestic and transnational politicisation. It highlights the role that these states play in international development[2] and the emergence of competing citizen expectations about what this role could and should be.

The creation of a new international financial institution (IFI) like the NDB is a symbolic component of a larger movement taking place since the early 2000s, whereby countries like the BRICS and other Southern middle-powers (including Mexico, Turkey and South Korea) became 'rising powers in international development' (Gu et al. 2016: 2). By championing 'alternative' development models and international development paradigms under the umbrella of South-South Cooperation (SSC),[3] these countries have played an increasingly greater role in shaping international development cooperation ideas, norms and practices and have contributed to significantly altering the geographies of development cooperation globally (Mawdsley 2012). The three countries discussed in this book – namely India, Brazil, and South Africa (also referred to as IBSA countries) – have been leading actors in championing SSC globally. At the same time, SSC has also grown in terms of political importance in the last decades within IBSA countries, generating new forms of domestic politics related to their international roles.

This chapter uses the case of growing mobilisation around the BRICS-led NDB from organised civil society groups in India, Brazil and South Africa to explore current domestic politics influencing foreign policy and the way these politics shape (and are shaped by) unfolding processes of value formation. Looking at foreign policy offers a less explored domestic policy domain[4] to discuss value formation in three postcolonial societies. The foreign policies of India, Brazil and South Africa are particularly interesting to explore through the lenses and the provocations offered in this book because they sit at the crossroads of two apparently contradictory but interrelated dynamics. On the one hand, as examined in several chapters in this volume (see for instance Chapter 3 by Cooper-Knock, Chapter 4 by Anciano, Chapter 7 by Mohanty and Chapter 9 by Schattan Coelho), as postcolonial societies, they exhibit particular sets of state-society dynamics characterised by increased frustration with fractured and failed promises made in the aftermath of particularly symbolic re-foundational moments (Indian independence in the late 1940s, the end of apartheid in South Africa in the mid-1990s, re-democratisation in Brazil in the late 1980s).

On the other hand, IBSA countries have also achieved considerable 'success' in overcoming a range of developmental issues through endogenous policy struggles that generated homegrown political compromises and policy solutions. These achievements have prompted international and peer recognition; transforming India, Brazil and South Africa into 'role-models' and sources of 'best practices' for others in the South (Gu et al. 2016; Stone et al. 2020). Backed by their relative socioeconomic successes and growing political will to act internationally on 'global public goods' in the realms of security, humanitarianism, environmental and developmental issues; their ambitious reformist diplomatic stances (including the creation of the NDB) have been permeated by both status-seeking counter-hegemonic narratives (Lopes et al. 2020) as well as global equality promises of power and wealth redistribution (van der Westhuizen 2012). Navigating between these two dynamics, their foreign policies are therefore marked by dilemmas and tensions generated from these global stances and their domestic implications (van der Westhuizen and Milani 2019; Lopes et al. 2020).

In this particular critical juncture, politicisation of foreign policy and development cooperation policies 'from below' reflects the multiple negotiations and 'social bargains' taking place in the marketplace (Siméant 2015) between states and organised citizens who navigate this sea of seemingly contradictory promises – some broken, some achieved and some that are yet to be fulfilled by now internationally recognised 'middle' or 'rising' powers. This is the context where social mobilisation around IBSA countries' foreign policies is increasingly taking place, manifesting plural understandings of national and transnational justice. In the past decade, domestic and cross-regional mobilisation around foreign policy and development cooperation issues has grown inside the three countries and in so-called transnational policy spaces, leaving behind what was seen as initial days of 'citizen apathy' on foreign policy and development cooperation issues (Mawdsley and Roychoudhury 2016; Pomeroy et al. 2016; Poskitt et al. 2016).

This chapter is an exploratory and circumscribed exercise on this transnational and multi-level research topic. Using the case of the BRICS-led NDB as a site of contention, it seeks to map emerging, and sometimes coexisting, political values shaping voice, participation and protest related to India, Brazil and South Africa's development cooperation. By doing so, it unpacks ongoing negotiations and disputes between state institutions and citizens, and more specifically organised civil society, regarding the role played by these countries in global affairs in the field of development cooperation and SSC. In this chapter, I argue that the intensification of collective action and citizen mobilisation – engaging, protesting and resisting development cooperation policies and practices – represents the politicisation of this agenda domestically and reflects a possible shift in citizens' expectations around foreign policy in countries that are simultaneously postcolonial and 'rising power' states. I also show how organised citizens' emerging political values are shaped by conflicting expectations of postcolonial states' development actions at home and beyond borders. Finally, I argue that while engaging with the external actions of these states, civil society groups in India, Brazil and South Africa bring to the transnational arena their multiple domestic political values and the political possibilities of voice and direct action. As such, political values and their manifestations are shaped by the different political structures and foundational agreements in each of these countries; as well as by the material, political and symbolic opportunities offered by national and transnational spaces.

In what follows, I introduce the case of citizen engagement with BRICS and their NDB. In doing so, I unpack the plurality of values shaping emerging forms of domestic and transnational citizen voice and action on this topic. Drawing on this case, I then offer a tentative conceptualisation of value formation in the context of postcolonial states' foreign policies and development cooperation policies. I conclude this chapter by discussing the potentialities of looking at value formation for understanding social mobilisation and foreign policy in postcolonial societies.

Value plurality in international development cooperation: Participation and protest in the new development bank

The BRICS grouping was forged in the early 2000s as a symbol of the nascent multi-polar post-Western world (Carey and Xiaoyun 2016). Yet, as the years went by, the economic growth underpinning most of the initial optimism around this grouping has either slowed down, stalled or completely vanished in several countries. Today, only China and India still carry that initial impetus, and in both cases, within more challenging political and economic environments. The very idea of investing in this geopolitical project became more contested domestically within each of the BRICS countries. However, in spite of the economic downturns and the severe domestic turmoil in countries like Brazil and South Africa in the field of international development, the grouping has actually succeeded in setting up the New Development Bank (NDB). This Shanghai-based multilateral development bank, created in 2014, is geared towards mobilising resources to finance sustainable infrastructure in the developing world. As of December 2019, the NDB has over forty projects approved in all five BRICS countries, with a total portfolio of approximately $15.1 billion in the areas of renewable energy, transportation and social infrastructure.[5]

Since the BRICS first announced their intention to create a bank of their own, the prospects of a 'new 21st century IFI' have sparked intense discussions within academic, policymaking and civil society spaces. Expectations were raised regarding the potential that the future BRICS-led institution had to 'change the rules and practices' of international development finance and to 'better serve' the needs of the developing South through alternative and sustainable development models (Waisbich and Borges 2019). Unsurprisingly, no consensus was reached regarding what 'alternative' or 'sustainable development' models to follow, or for that matter, on what NDB's 'sustainable infrastructure' mandate meant and to whom.

Even before the formal start of NDB's operations, the bank had galvanised attention from already mobilised civil society groups working either on BRICS-related affairs or on development finance more broadly. For these groups, there was a 'unique window of opportunity' to influence the very set-up of the new institution considering that: (i) the NDB is the first concrete outcome of the BRICS partnership; (ii) the NDB is the first IFI created by developing countries to serve developing countries; (iii) the NDB had an upfront intention of being an innovative niche bank funding sustainable infrastructure (Waisbich 2017).

NDB's operations started in 2016, but most of its funded projects took a few years to take off, rendering NDB's activities quite discreet until the present moment. The little materiality and visibility of projects on the ground has, on the one hand, hindered the rise of more conventional community-based mobilisation (Fox and Brown 1998), namely transnational campaigns and on-site protests by those who consider themselves as 'negatively affected' by NDB projects. On the other hand, it has also provided a small group of civil society actors – mostly professionalised non-governmental organisation (NGOs) from IBSA countries and, to a lesser extent, China – the opportunity to engage the bank in a more policy dialogue fashion.

Voice, participation and protest in BRICS summits

Civil society engagement with the broader BRICS agenda has been context-based and shaped by the particular state-society dynamics playing out in each country, as well as the degree of openness to civil society dialogue across the five member states. It has also been shaped by the specific mobilisation opportunities arising from the nature of BRICS as a grouping, particularly its rather abstract 'reforming global governance' agenda and its rotating annual BRICS heads-of-states summits (Pomeroy et al. 2016; Poskitt et al. 2016; Waisbich 2017; Thompson and De Wet 2018). Patterns of mobilisation have combined, on the one hand, continuous low-intensity activism by a handful of think tanks and a larger, but still relatively niche, group of NGOs and social movements – mostly from the IBSA countries[6] – with a clear international agenda and resources to engage on foreign policy issues and take part in international negotiations. On the other hand, more intense mobilisation also took place in the lead-up to (or during) the annual official BRICS summits taking place in one of the five member states.

Since 2013, civil society engagement with the BRICS agenda has grown considerably and civil society groups have been successful in mobilising and carving out modest, yet sustained, spaces for policy dialogue through workshops and seminars; or mobilising at the margins of official gatherings, including through counter-summits (Waisbich 2017). Consequently, a myriad of 'invited spaces' and 'claimed spaces' – to use John Gaventa's (2006: 26) typology on participation spaces – have mushroomed at the national level (again, mostly in IBSA countries) as well as at the sidelines of the annual official BRICS heads-of-states summits. As mobilisation grew at the international level, two types of participation spaces mirroring the dynamics of insider-outsider civil society activism started to co-exist: on the one hand, the independent counter-summit events, also referred to as the *People's BRICS* spaces. On the other, the *Civil BRICS*, comprised of semi-official gatherings partially integrated into the official BRICS summit agenda and negotiated with the hosting government.

The very first civil society gathering, named *BRICS-From-Below* by the South African activists leading it, took place in Durban in 2013 and was followed by a similar one in Brazil in 2014. In 2015, Russia decided to implement the *Civil BRICS* model, mirroring the format for citizen participation adopted in G20 meetings. As an invited space, the 'civic/civilised gathering' got unprecedented official recognition in the BRICS summit official agenda, but brought together less critical voices from NGOs and social movements. In India in 2016, two types of events were held and took place almost simultaneously, albeit in different cities (i.e. Delhi for the *Civil* BRICS, Goa for the *People's* BRICS). The following year in China, the main event was the official government-sanctioned *Civil* BRICS attended by think tanks and parliamentarians mostly, with a small and rather marginal labour union-led counter-summit taking place in Hong Kong. In 2018, South Africa presided over the BRICS summit and once more, mobilised groups divided themselves into invited (*Civil*) and claimed (*People's*) spaces, with the latter also organising the anti-NDB street demonstrations in Johannesburg. Finally, during the Brazilian presidency of 2019, there was no major civil society event at the margins of the official summit, although some civil society groups and left-wing politicians (formerly in government, and now in opposition to Jair Bolsonaro's presidency)[7] had organised small BRICS-related debates.

Autonomous or claimed civil society-led *People's* spaces have successfully assembled a range of national organisations and movements whose activism was either intimate, direct or even indirectly linked to the BRICS agenda. Among those who hosted and/or attended the counter-summits, there were emblematic national social movements (including landless workers in Brazil, mine workers in South Africa and environmental groups fighting against nuclear and coal plants in India). In each counter-summit, these social movements were joined by several other local groups who presented themselves as 'BRICS-affected peoples/communities' (or 'negatively impacted by BRICS'), including: groups affected by the South Durban Port (2013), groups affected by World Cup-related forced evictions in Fortaleza (2014), and groups affected by predatory tourism in Goa (2016). Making use of the international spotlight, these local groups have used the opportunity of having their countries/cities officially hosting a BRICS event to amplify their domestic struggles and find international solidarity abroad around the notion of being 'affected by BRICS'. The participation of local groups was mostly once-off, since they rarely took part in subsequent counter-summits, making it difficult to establish a continuous thematic agenda for *People's* summits throughout the years.

Although sometimes less connected to the BRICS agenda per se, their voices illustrate some of the most pressing social issues – or the broken or allegedly broken promises and agreements – of each of the three postcolonial states, at both national and subnational levels. Their presence has also been key to giving a concrete face to frequently advanced, but often vague, organised civil society narratives and concerns with 'BRICS's exclusive or predatory developmental models' beyond borders. The plight of local groups illustrates, therefore, the negative socioeconomic and environmental impacts of 'unfair' and 'exploitative' development models on the ground for marginalised communities within BRICS countries. On the other hand, giving visibility to prevailing domestic social injustice fits the more professionalised groups' strategy to counter what they see as global social injustice fostered by the BRICS countries through their foreign policies and/or their investments in third world countries through instruments like the NDB.

The diversity seen in participation spaces (both between *Civil* and *People's* spaces, and among the different types of civil society groups that joined them) also mirrors different types of political values in the making, representing diverging expectations on states' development models and international development engagements, as well as expectations on citizen participation and state-society interfaces. Scholars such as Patrick Bond and Ana Garcia (2015) have proposed a typology for BRICS civil society actors. This typology is made up of two groups: the *BRICS-from-Below* (comprised of more radical – often left-wing – oppositional voices from social movements) and *BRICS-from-the-Middle* (the more conforming and less critical voices coming from professionalised, advocacy-based non-governmental organisations). This categorisation highlights the degree to which some social groups share affinities with official actors, narratives, policies and practices; and how these affinities determine civil society groups' potential access to policymakers. Using Bond and Garcia's (2015) analysis, and drawing on Cooper-Knock's (in this volume) 'politics of patience' framework to study moral economy in post-apartheid South Africa, I contend that the divisions between *From-Below* and *From-the-Middle* (or *People's* versus *Civil* BRICS spaces) also reveal the multiple degrees of 'patience' in terms of the so-called 'transformative' politics advanced by BRICS nations in their collective 'anti-systemic' and 'revisionist' international developmental endeavours.

However, it is important not to over-essentialize these categories and to pay more attention to hybrid identities that shape value formation and engagement with authorities (using strategies ranging from dialogue and advocacy to boycotts). It is equally important to investigate how these categories have been changing throughout the years in successive official summits and within countries – thus reflecting political possibilities and political opportunities for foreign policy/BRICS-related voice and action (Waisbich 2017). For instance, the division between 'civilised' and 'unruly' – with their insider-outsider strategies or 'tree-shakers' and 'jam-makers' strategies (Mdlalose and Thompson 2018) – have been particularly visible in South Africa and India, whereas Brazilian groups have been significantly less fractured. Likewise, street protests against the BRICS have been utilised more as an action repertoire in South Africa than in the remaining countries.[8] Looking at the mobilisation around the NDB hence offers an opportunity for exploring some of these nuances; while also examining the formation of identities, values and repertoires of collective action among mobilised groups, as I will show in the next section.

Voice, participation, and protest around the New Development Bank

In this section, I look at the mobilisation repertoires of civil society groups when engaging with the BRICS-led NDB. Using discourse analysis, I examine advocacy documents published in 2017 by two different cross-regional civil society coalitions monitoring the NDB. These two cross-regional civil society coalitions are comprised of members from all the BRICS countries and led by organisations based in India, Brazil and South Africa. Not only does this documentation on advocacy illustrate discursive battles over tactical dimensions of collective action, but it also reveals the formation of political values in relation to perceptions of social justice, 'good foreign policy' and 'good/responsible international state behaviour'.

The first document, titled *The BRICS New Development Bank Strategy: A Civil Society Perspective for Truly Sustainable Infrastructure and Transformative Development Cooperation*, is a policy report, crafted collectively by the Coalition for Human Rights in Development. The Coalition is a transnational umbrella-organisation working on IFIs, whose Working Group on BRICS gathers NGOs from the BRICS countries and also some of the biggest development INGOs.[9] Organisations have been monitoring the NDB since its inception and have jointly articulated policy recommendations for the bank. The document was produced to influence the NDB 2017–2021 Strategy, the first official programmatic document released by the bank. In this policy report, organisations raise a series of socio-environmental and accountability-related demands to the NDB, requesting that the bank prioritise investments 'in both social and environmentally sustainable infrastructure', 'incorporate global best practices on social and environmental management, transparency, and accountability' and formulate 'institutionalized mechanisms for meaningful engagement with civil society and develop clear sustainability criteria and performance targets for its investments' (Coalition for Human Rights in Development 2017: 5). Their expertise with traditional IFIs and their human rights-based framings for NDB accountability include demands for the bank to create project-level accountability mechanisms for consulting with affected communities and an independent accountability mechanism to 'provide redress for individuals or communities which may be impacted by NDB-financed activities, with powers to make binding decisions and sufficient technical, financial and human resources to evaluate the NDB's operations against its own standards' (Coalition 2017: 12). The language in this first document is very consistent with the rights-based international advocacy claims historically advanced towards most of the traditional IFIs, and reveals a particular set of 'do-no-harm' expectations on the BRICS-led NDB.

The second document, titled *New Development Bank – Peoples' Perspectives* (People's BRICS 2017), is a public statement. It was drafted by a group of mostly South African and Indian self-identified 'radical and grassroots organisations' working under the umbrella of the *People's Forum on BRICS*. The document was issued in March 2017, in the context of NDB's second annual meeting in Delhi. The statement makes use of less 'consensualising and conflict avoiding jargon language' – terms borrowed from Swyngedouw (2013: 826) – and criticises the bank for being 'unaccountable', and not 'having 'pro-actively reach[ed] out to peoples movements, trade unions, civil society organisations and peoples networks genuinely working with poor and impacted communities' (People's BRICS 2017). In their communiqué, the signatories also criticise the bank for 'finding comfort in plush and polite consultations held with fellow bankers, bureaucrats, key politicians and cherry-picking "NGOs" who conform with rather than critically engage with the NDB' (People's BRICS 2017). This second piece reveals an alternative take on the relationship between civil society voices and the bank. The document emphasises the idea of 'people's voices', underlining signatories' identity as organisations fighting injustice domestically in the BRICS and accusing the NDB of 'promoting a business as usual model, rather than striking a dramatically new path for people-centred development that was promised in the formation of the Bank by the BRICS nations'. For this group, the bank suffers from 'systemic myopia', promotes rather 'hollow' models of sustainability, and is not responsive to 'people's real concerns' (People's BRICS 2017). Ideas around social justice and people-centred development, respect for labour conditions, freedom of association and the protection of the environment are at the core of People's BRICS NDB mobilisation and of their calls and expectations about what kind of projects the bank should be funding.

There are some striking commonalities as well as some important divergences between the two documents. Both emphasise participation demands, suggesting ways in which the bank should engage with civil society and whom in civil society should be prioritised. Both documents also challenge NDB's current institutional architecture and policies, with strong criticism directed towards what they see as an elitist approach to development by the bank and little concern for the marginal or vulnerable communities affected by its projects. Civil society groups are also vocal in pointing out the gaps between NDB's sustainable or 'green/sustainability rhetoric' and its funding practices. Considering NDB's portfolio of more than forty approved projects so far, several of them labelled as 'sustainable infrastructure' projects, groups have expressed their concerns with the overall 'developmental impact' of NDB's projects. When criticising NDB's 'transparency and accountability' gaps – including the lack of publicly available criteria for what the bank understands by the term 'sustainable infrastructure',[10] and the risks of 'green washing' exemplified by NDB's unwillingness to adopt a more ambitious low-carbon funding strategy or a gender policy – active civil society groups not only mobilise particular conceptions of development and international development, but also bring to the public arena expectations on BRICS states' development finance and development cooperation to adhere to those conceptions.

The critical engagement of a range of civil society groups from IBSA countries with the NDB, and the way they frame their criticism by pointing to the perceived gaps between the diplomatic 'reformist' rhetoric of BRICS as a political grouping and the institutional practices of the NDB (arguably BRICS's first and most concrete institutional creation) also reveals existing state-society contentions and the emerging breakdown of (tacit) initial agreements around this international role by the grouping, perceived as falling short of the social-environmental justice expectations held by mobilised civil society groups.

Value plurality regarding IBSA countries' international development cooperation?

This brief discourse analysis reveals multiple framings shaping participation and protest by mobilised groups in India, Brazil and South Africa on BRICS and NDB issues. It shows mobilised groups politically engaging with imagined or perceived NDB policy/operations outcomes as much as with the bank's performance in its technical-managerial dimension (including its social-environmental risk and impact assessment procedures). Both groups make use of discursive strategies and storylines connecting struggles across countries (and also between domestic struggles and transnational ones), particularly on issues related to environmental damage and transparency. Yet, while some groups would radically question NDB's policy options and policymaking processes, others act instead as insiders: forging alliances to push softer 'pro-poor or rights-based' reforms from within.

From a tactical perspective, this insider-outsider divide does create challenges to collective action and to the formation of a broader critical mass of civil society actors capable of putting pressure on the NDB. However, in what concerns a central theme of this edited collection – namely the relationship between political values and resistance – this divide is equally revealing of the diversity within civil society regarding claims on postcolonial states' role as development actors, and coexisting values and expectations on the political possibilities to engage state institutions and shape their foreign policies/development cooperation. These distinctions are embedded in different values and conceptions of social justice, including the very value of engaging the state and state institutions to influence its policies and practices. In other words, the value of voice through institutionalised participation. This is clear when some of the more 'patient' groups, who see themselves as 'NDB stakeholders', opt for insider strategies and constructive engagement with the NDB representatives, propose meetings and take part in NDB events. Others, who frame their work more along the lines of 'peoples-affected by the NDB', opt for outsider strategies like the *People's Audits* discussion panels in India, the street protests in South Africa or even public silence, as in the case of the very weak public mobilisation in Russia and China (on silence, see Ngwenya and von Lieres, Chapter 2, this volume).

From a participation perspective, it is fair to say that both groups still face classic representation challenges. The connections between those who can actually engage with global development debates and/or the bank and their staff in Shanghai in a highly specialised technical language and in English, and local communities impacted by the projects they claim to represent, remains incipient. So far, advocacy has taken place abstractly – targeting NDB's operational policies or expected portfolio-building strategies – with little reference to particular communities or projects, except one road project in India (BRICS Feminist Watch 2019) and one port expansion in South Africa. Most of the activists have no permanent links to the communities affected by NDB projects, except the Durban port case where a continuous grassroot mobilisation linking this infrastructure project to global investments has been taking place for several years now (Bond and Mottiar 2018), which can explain why South Africa was the only country where anti-NDB street demonstrations were held so far. For the remaining projects, one can expect development and development finance experts, or even leaders of nation-wide social or labour movements, to develop *ad hoc*, limited and time-bounded linkages with affected communities.

Ultimately, the very value of a 'right not to be negatively affected by developmental projects' has a very difficult collective action translation in the case of the NDB. Mobilisation is challenging for, at least, two reasons. First, the bank has been playing safe and has chosen to start with projects with little visible material impact on the ground (Waisbich and Borges 2019). Second, NDB's business model implies a real challenge for attribution and responsibility, since it funds projects through national intermediaries (including national development banks, public agencies and the private sector) and uses the 'country systems' as the main framework for public financial management, public contracts, audits, monitoring and assessments, and social and environmental procedures (NDB 2017). The country system approach makes it particularly difficult for activists to apply classic strategies of 'project-tracing' and 'naming-and-shaming', historically quite successful in the case of projects funded by the World Bank, including in countries like India and Brazil (Fox and Brown 1998). As such, NDB accountability politics depend even more on the domestic politics of citizenship and participation of its members; and on the political opportunities they find at home not only to challenge NDB's operations but also their countries' foreign and domestic development policies.

Finally, there is also value plurality among mobilised groups on what kind of 'just developmental role', if any, postcolonial states should play – including whether or not to adopt a developmentalist state approach, and what conceptions of development/development cooperation would shape states' and NDB's interventions. Divergent takes on 'sustainability' and the role of the private sector (and public–private partnerships) in development finance are the most visible contentious issues, but not the only ones. Civil society groups also hold competing views on what 'beyond growth models' or 'environmental sustainability' means. While some groups would embrace 'Green Economy' narratives and projects, others would prefer more openly 'pro-poor', 'people-centred' or 'women-led' initiatives, mobilising those particular understandings of social justice and social good policies to refer to the state role in development domestically and abroad.

Value formation and the politicisation of development cooperation in IBSA countries

Civil society transnational mobilisation in the context of the BRICS-led NDB offers valuable insights to discuss political value formation related to postcolonial developmental states' foreign policies. In this section, I discuss a tentative conceptual lens into this phenomenon, bridging the processes of value formation and domestic politics of foreign policy in the context of growing development cooperation in India, Brazil and South Africa.

My initial assertion is that civil society voice, participation and protest dynamics – such as those described in BRICS/NDB-related mobilisation – are to be understood within the growing politicisation of foreign policy matters in postcolonial societies, a phenomenon scholars also relate to an ongoing process of 'foreign policy democratisation' since the 1990s (Sanchez et al. 2006; Lopes 2010; Nel and Van der Westhuizen 2003). In this context, increased civil society mobilisation is shaped by existing – and also competing – values, social understandings and expectations around public and social goods and how they fit state actions at home and abroad, including through foreign policy acts and deeds.

As an emerging policy field within the broader foreign policy realm, development cooperation is also subjected to and shaped by a set of values, understandings and expectations related to what is deemed appropriate behaviour for postcolonial states when it comes to promoting development (or fighting poverty) at home and abroad. Is development an (international) public good? Is it about international solidarity and global justice, or about fostering good relationships that can procure material and symbolic gains in the future? What specific sectors and issues should be prioritised (if any)? How much of the national budget should be devoted to those policies? Should states mobilise domestic resources to fund international poverty alleviation, infrastructure building or trade promotion?

Domestic politics of foreign policy and value formation

Foreign policies can be insightful sites to investigate value formation and its manifestations in postcolonial developmental states. Even if less politically salient than many other realms of public life (Hill 2003; van der Westhuizen and Milani 2019), foreign policies directly and indirectly affect citizens' lives within national borders as well as the lives of citizens in other jurisdictions. With greater political and economic liberalisation since the 1990s, the foreign policies of countries in the 'Global South' have also become more distributive (Lima 2000) and are thus subjected to sociopolitical bargaining among a range of different domestic stakeholders over their material, political and symbolic returns (Sanchez et al. 2006; Lopes 2010; Alden and Brummer 2019).

The domestic politics of foreign policy offer a promising analytical framework to assess the emerging political values, preferences and repertoires of collective action they shape, such as those found in the case of the BRICS/NDB mobilisation. Hill (2003: 23) defines the politics of foreign policy as 'the problem of acting in international affairs, through state and other actors, and of balancing the competing pressures and expectations which beset any foreign-policy maker'. Competing pressures, he adds, can also be seen as 'a simultaneous stream of internal and external demands upon government' (Hill 2003: 23), which means competing claims and multiple responsibilities now facing decisionmakers. At the domestic level, competing pressures on those responsible for foreign policy can be seen not only through the lens of intra-state or bureaucratic conflicts (in the classic bureaucratic politics framework), but also through the lens of state-society relations (Alden and Brummer 2019).

Even though Hill's (2003) overall argument is based on a collection of examples gathered from the so-called 'industrialised' democracies, mostly in the West, his proposition also suits postcolonial democracies like India, Brazil and South Africa. Working with these countries enriches Hill's discussion with a different set of institutional and sociopolitical dynamics found in postcolonial settings, most notably the tensions generated by the simultaneity of broken and renewed promises.

This gaze into the distributive conflicts and the disputed nature of foreign policy and development cooperation/SSC policy in IBSA countries is an important analytical tool to investigate the citizenship-related questions treated in this book, in particular questions around the formation of political values in relation to perceptions of social justice; as well as how political values manifest in different modes of engagement with public authority in foreign policy-related matters. In this regard, the politicisation of foreign policy observed in the past decades in all three countries can be understood as generative of political subjectivities and as an enabler of value formation among citizens and societal forces to question and dispute the making, enacting and effects of foreign policy as a public policy. As such, disputes and claims 'from below' (from citizens and/or organised civil society) related to foreign policy issues rely on explicit, stable and even emerging agreements among state and societal forces around the right to question, participate and/or demand accountability on foreign policy issues.

As suggested by Hossain and Kalita (2014: 825), 'in multiparty democracies, however flawed, people feel some power over policymakers and decision-takers'. How, then, to make sense of emerging 'feelings' concerning IBSA foreign policies: are they driven by the same domestic values shaping state-society interactions on other policies? How applicable are those foundational social contracts ruling domestic policies and politics to foreign policies? How do the transnational opportunity structures, reflecting strong sovereignty and non-interference principles underpinning contemporary international relations and postcolonial states development cooperation, shape not only those values but also the possibilities of contentious collective action?

In this chapter, I offer three contributions to understanding the links between the contested nature of foreign policy making in postcolonial settings and value formation. First, engaging with Isunza and Lavalle's (2018) idea of the 'right to demand accounts' (what they call in Spanish the *exigibilidad*) and looking at the politics of foreign policy in time, I posit that value formation around foreign policy in IBSA countries in the current moment of their historical and sociopolitical trajectories is shaped by two coexisting sets of emerging values, formed by two distinct and mutually reinforcing scales of analysis. The first one, the *local scale*, is the one where the domestic distributive effects of foreign policy are more immediately thought and felt by organised citizen groups and is thus related to current legitimizing notions of national justice, the role of foreign policy as a state policy, and its direct effects on people's lives. The second one, the *transnational scale*, relates to the cosmopolitan lens of global justice and to legitimizing notions of states and societies acting internationally as part of a bigger interdependent community, and acting globally to secure development across national borders. The second contribution speaks to the types of values being formed, which in the case of foreign policy/development cooperation relate to notions of the good life, good policies, and good states (including notions on whether or not to have a 'developmental state'), as well as good policymaking. My final contribution, drawing on the case of the NDB-related social mobilisation described above, is that multiple competing values are being formed and inform different foreign policy preferences and different modes of engagement with public authority. Diversity, even within civil society, responds to the current polarised policy and political arenas regarding broken promises/rising power status, opposing international activism and national developmental challenges at home (Mawdsley and Roychoudhury 2016; UNDP China 2018; van der Westhuizen and Milani 2019), but also traditional political and partisan divides around the role states (or the public sector) should play in development.

Development cooperation/South-South cooperation domestic 'constituencies'

Value plurality and its manifestations can be further understood in dialogue with the notion of domestic foreign policy 'constituencies' or 'development cooperation constituencies', present in Hill's (2003) work as well as in more specific studies on 'traditional development cooperation/foreign aid' (Lancaster 2007). Domestic constituencies include lawmakers, development academics, CSOs, the private sector, as well as the so-called general public and printed opinion. Neither foreign policy constituencies nor development cooperation/SSC constituencies are monolithic. Moreover, SSC scholars have come to the conclusion that countries like India, Brazil, and South Africa have a complex, if not tense, relationship with their multiple constituencies due to the ambivalences, dilemmas and traps attached to their position in the international system as middle/rising powers (van der Westhuizen and Milani 2019). These traps are both global and domestic in nature. Globally, middle-powers are not easily welcomed and unavoidably collide with existing powers and their institutions due to their reformist/revisionist aspirations (Lopes et al. 2020). Domestically, dilemmas can be of a distributive nature, with different national constituencies expecting national benefits in return for international activism, but also advancing competing ideas and values around social justice and just policymaking.

Dilemmas are particularly acute in development cooperation/SSC due to a range of factors. First, the low-level of consolidation of this policy field within IBSA countries (Mawdsley 2019; van der Westhuizen and Milani 2019). Second, the material and political limitations posed by domestic needs (namely fighting poverty and inequality at home) to sustaining public support and engagement with other developing countries (Six 2009; UNDP China 2018), particularly those generating, or perceived as generating, greater financial commitments (Hardt et al. 2020). Third, the SSC's own geopolitical rationale. Framed as 'win-win' or 'mutual benefits', rather than 'aid', 'charity' or 'gift' (Mawdsley 2012), SSC has from the start been a political-normative endeavour open to disputes and bargains by policymakers and citizens alike, at least in vibrant democracies such as India, Brazil and South Africa. As a normatively charged project, development cooperation by IBSA countries is shaped by constantly shifting expectations on their international development policies and their outcomes, as much as their national and global responsibilities.

Conclusion

This chapter explores the formation of values and their manifestation in terms of voice, participation and protest regarding international development cooperation matters in postcolonial societies like India, Brazil and South Africa. It looks at growing attempts by organised civil society groups from these countries to influence and challenge the BRICS-led New Development Bank (NDB) and unpacks the ways in which different groups articulate their demands. The chapter further examines the politicisation of development cooperation or South-South cooperation for development (SSC) and, more broadly foreign policy, from below.

It shows how political values shape citizen groups and civil society organisation attitudes towards public authorities in the case of NDB-related mobilisation; and values that relate to both citizens' notions of the good life, such as social and cosmopolitan justice, as well as good policymaking – including transparency and participation. The chapter also highlights the plurality of values and ideas around citizenship and development underpinning emerging conceptions of citizen participation in postcolonial states' international development engagements and the ongoing negotiations resulting from those diverging understandings. These disagreements, even in their low-intensity format, are not only a window into the politicisation of development cooperation in postcolonial states like India, Brazil and South Africa; but are also revealing of dynamics of value formation and representations of social justice and of 'pro-poor' or 'rights-based' development shaping expectations on states' roles in promoting development at home and abroad. The case of civil society mobilisation around the NDB exposes fractures in the short-lived agreement between BRICS states' narratives on promoting international justice (or democratising global governance) and the narratives advanced by civil society groups, focusing rather on global people-centred justice (at home and beyond borders).

A closer look at this particular transnational civil society mobilisation illustrates recent forms of 'development politics' through the examination of social justice claims cutting across national and international realms of state action. Those claims are embedded in symbolic, discursive and material battles over who gains and who loses in development processes, whose voice counts to negotiate what development and development cooperation is or should be, and – ultimately – it questions how foreign policies are negotiated in certain parts of the 'rising South' and how those policies affect, reinforce and are shaped by 'how citizenship is exercised, enforced and denied' in those same places (Newell and Wheeler 2006: 50). Informed by the broader participation and transnational social movements literature, this chapter has also touched on the domestic and transnational political opportunity structures bounding the strategic choice of voice, participation and protest, illustrated in the case of BRICS-led NDB by insider-outsider engagements. How organised citizens frame their protests depends on the politics of each particular policy arena and, in this case, on the politics of multiple arenas acting simultaneously at different and interrelated scales. This multi-scalarity influences how values and norms play out in the different postcolonial states' development cooperation engagements and in their collective enterprise: the NDB.

Ultimately, this exploratory reflection on civil society reactions to the NDB contributes to recent and growing attempts to bridge critical foreign policy analysis and broader citizenship literature through the investigation of value formation in postcolonial states' international development cooperation/SSC. Looking at the domestic politics of foreign policy thus allows for unpacking how citizen groups in postcolonial settings understand foreign policy decisions and their instruments, including the BRICS bank. This combined take is a useful heuristic device to illuminate the emerging societal contestation by citizens and organised civil society in India, Brazil and South Africa around their countries' impact beyond borders, as well as ongoing disputes over the multiple meanings of development and development cooperation (*what kind of development?*, *who benefits from development interventions?)*. The state-society focus presented here also contributes to the growing body of work on the domestic politics of SSC (Leite et al. 2014; Cabral 2016; Mawdsley and Roychoudhury 2016), reiterating the calls for a deeper societal lens into the pluralisation – or even democratisation – of foreign policy-making processes and how are they shaped by processes where citizenship itself is being constructed and contested in tandem with durable inequalities (Von Lieres and Piper 2014). Such a power-sensitive analysis of how different non-state actors engage with development cooperation, within the broader umbrella of foreign policy, highlights the disputed nature of foreign policy and SSC debates, forums and mechanisms in Brazil, India and South Africa. It also contributes to a better understanding of foreign and development policies and practices, while questioning the normative implications for state and democratic legitimacy in IBSA countries – and more broadly, for emerging forms of globally active citizenship in the Global South.

Finally, this chapter also leaves some open avenues for future research on domestic value formation and foreign policy in postcolonial states. The present cross-country study focused primarily on value formation and their manifestations using the examples of highly organised and internationalised civil society groups acting transnationally, many of them belonging to elite groups in Brazil, India, and South Africa. Better understanding popular notions of legitimacy and social justice across a wider range of societal actors in each of those countries and the extent to which they converge or diverge with those presented here is thus a promising research agenda. Some recent work on public opinion and foreign aid in Brazil and South Africa (van der Westhuizen 2012; Hardt et al. 2020) provide insights on how regional, class and age diversity manifest in foreign policy/development cooperation issues, but this research agenda could be further enriched by studies employing methodologies other than opinion surveys. Future work on the topic can also highlight new dynamics and potential value-clashes taking place in IBSA countries as a result of increased political polarisation domestically.

Notes

1 Examples of the coverage, in the South Africa media, of the 2018 demonstration can be seen at: https://citizen.co.za/news/south-africa/1987116/break-the-brics-protesters-march-outside-the-summit/. For the 2019 demonstration, see https://dialogochino.net/25543-ndb-slow-to-deliver-on-promises/.

2 Other than the term 'postcolonial states' used in this collection, scholars studying the rise of the BRICS in international development and the re(emergence) of South-South Cooperation (a modality of development cooperation they champion) have also referred to these countries as 'new donors', 'new development cooperation providers', 'rising powers in international development', and 'non-Western or non-DAC donors'. For an explicit use of the term 'postcolonial states' to refer to the so-called Eastern and Southern donors, see Six (2009).

3 According to the United Nations Office for South-South Cooperation (add page number), SSC can be defined as 'an exchange of knowledge and resources in the political, economic, social, cultural, environmental or technical domain between developing countries. It can take place on a bilateral, regional, sub-regional or interregional basis and can involve two or more developing countries'.

4 Here I used those terms interchangeably to denote, as Burstein (1991: 328) once described, 'the components of political systems organized around substantive issues'.

5 A list of all projects approved by NDB's Board of Directors can be retrieved at https://www.ndb.int/projects/.

6 On the side of Russia and China, participation has been mostly dominated by think-thanks, with some environmental NGOs from China also being increasingly involved, particularly since the creation of the NDB.

7 This was the first time that high-profiled politicians, mostly Congressmen, have joined civil society-led counter-summit events. Although an in-depth analysis of the political dynamics in Brazil escape from the scope of this chapter, it is important to highlight that the 2019 summit in Brazil clearly marks a turn in the pattern of Brazilian civil society engagement in BRICS-related events.

8 The landscape in IBSA countries also contrasted with the situation of groups in China and Russia, where critical voices willing to adopt an openly critical or confrontational tone have been extremely rare.

9 This includes Oxfam, International Rivers, Accountability Counsel and Friends of the Earth, to name a few.

10 At the time of the NDB Strategy release, in 2017, the Bank already had several projects approved and under operation; and neither the Strategy nor additional documents provided clarity on what the bank understood and conceived as 'development impact' or 'sustainability' and how it was planning to asses them. In the following two years, the bank made important advances on this front, with a clearer narrative of its own benchmarks and metrics of impact and sustainability.

References

Alden, C., and K. Brummer. 2019. 'Foreign Policy Analysis and the Study of Indian Foreign Policy: A Pathway for Theoretical Innovation?' *India Review* 18(5): 471–484.

Bond, P., and A. Garcia. 2015. *BRICS: An Anti-Capitalist Critique*. London: Pluto Press.

Bond, P., and S. Mottiar. 2018. 'Terrains of Civil and Uncivil Society in Post-Apartheid Durban', *Urban Forum* 29(4): 383–395.

BRICS Feminist Watch. 2019. *Gender Monitoring of New Development Bank Project: Major District Roads, Madhya Pradesh, India – Key Findings*. New Delhi: BRICS Feminist Watch/PWESCR.

Burstein, P. 1991. 'Policy Domains: Organization, Culture, and Policy Outcomes', *Annual Review of Sociology* 17(1): 327–350.

Cabral, L. 2016. 'Priests, Technicians and Traders: Actors, Interests and Discursive Politics in Brazil's Agricultural Development Cooperation Programmes with Mozambique', PhD Dissertation. Sussex: University of Sussex.

Carey, R., and L. Xiaoyun. 2016. 'Understanding the BRICS Evolving Influence and Role in Global Governance and Development', *IDS Policy Briefing Issue 119*. Brighton: Institute of Development Studies.

Coalition for Human Rights in Development. 2017. *The BRICS New Development Bank Strategy: A Civil Society Perspective for Truly Sustainable Infrastructure and Transformative Development Cooperation*. Last accessed 30 June 2020. https://rightsindevelopment.org/wp-content/uploads/2017/04/Folheto-The-BRICS-Sustainable2.pdf.

Fox, J., and L.D. Brown. 1998. *The Struggle for Accountability: The World Bank, NGOs, and Grassroots Movements*. Massachusetts: MIT Press.

Gaventa, J. 2006. 'Finding the Spaces for Change: A Power Analysis', *IDS Bulletin* 37: 23–33.

Gu, J., A. Shankland and A. Chenoy. 2016. *The BRICS in International Development*. Basingstoke: Palgrave Macmillan.

Hardt, M.S., F. Mouron and L.A. Júnior. 2020. 'Brazilian International Development Cooperation and Public Opinion: Domestic Costs Faced by a Troubled Emerging Donor', *Brazilian Political Science Review* 14(1): e0001.

Hill, C. 2003. *The Changing Politics of Foreign Policy*. Basingstoke, Hampshire: Palgrave Macmillan.

Hossain, N., and D. Kalita. 2014. 'Moral Economy in a Global Era: The Politics of Provisions during Contemporary Food Price Spikes', *Journal of Peasant Studies* 41(5): 815–831.

Isunza, E.V., and A. Gurza. 2018. *Controles Democráticos No Electorales y Regímenes de Rendición de Cuentas En El Sur Global*. Oxford: Peter Lang Publishing.

Lancaster, C. 2007. *Foreign Aid: Diplomacy, Development, Domestic Politics*. Chicago: University of Chicago Press.

Leite, I.C., et al. 2014. *Brazil's Engagement in International Development Cooperation: The State of the Debate*. Sussex: Institute of Development Studies.

Lima, M.R.S. 2000. 'Instituições Democráticas e Política Exterior', *Contexto Internacional* 22(2): 265–303.

Lopes, D.B. 2010. 'Da Razão de Estado ao Republicanismo Mitigado: A Plausibilidade de uma Política Externa Democraticamente Orientada no Brasil Contemporâneo', PhD Dissertation. IESP/UERJ.

Lopes, D.B., G. Casarões and F.C. Gama. 2020. 'A Tragedy of Middle Power Politics: Traps in Brazil's Quest for Institutional Revisionism', in P. Esteves, M.G. Jumbert and B. de Carvalho (eds.), *Status and the Rise of Brazil: Global Ambitions, Humanitarian*

Engagement and International Challenges. New York: Springer International Publishing, pp. 51–69.

Mawdsley, E. 2012. *From Recipients to Donors: Emerging Powers and the Changing Development Landscape*. London: Zed Books.

Mawdsley, E. 2019. 'South–South Cooperation 3.0? Managing the Consequences of Success in the Decade Ahead', *Oxford Development Studies* 47(3): 259–274.

Mawdsley, E., and S. Roychoudhury. 2016. 'Civil Society Organisations and Indian Development Assistance: Emerging Roles for Commentators, Collaborators, and Critics', in S. Chaturvedi and A. Mulakala (eds.), *India's Approach to Development Cooperation*. London: Routledge, pp. 79–93.

Mdlalose, B., and L. Thompson. 2018. 'Are BRICS Civil Society Talkshops just Ticking Boxes and Not Making Real "Jam"?'. *Mail & Guardian*. Last accessed 11 May 2019. https://mg.co.za/article/2018-06-06-are-brics-civil-society-talkshops-just-ticking-boxes-and-not-making-real-jam.

Nel, P., and J. Van der Westhuizen. 2003. *Democratizing Foreign Policy? Lessons from South Africa*. Lanham, MD: Lexington Books.

New Development Bank (NDB). 2017. *NDB's General Strategy 2017–2021*. Shanghai: New Development Bank.

Newell, P., and J. Wheeler. 2006. *Rights, Resources and the Politics of Accountability*. London: Zed Books.

People's BRICS. 2017. *New Delhi Declaration: New Development Bank – Peoples' Perspectives*. April 21, 2017. Last accessed 30 June 2020. https://publicservices.international/resources/news/new-delhi-declaration-new-development-bank--peoples-perspectives?id=7694&lang=en.

Pomeroy, M., et al. 2016. 'Civil Society, BRICS and International Development Cooperation: Perspectives from India, South Africa and Brazil', in J. Gu, A. Shankland and A. Chenoy (eds.), *The BRICS in International Development*. London: Palgrave Macmillan, pp. 169–206.

Poskitt, A., A. Shankland and K. Taela. 2016. 'Civil Society from the BRICS: Emerging Roles in the New International Development Landscape', *IDS Evidence Report 173*. Brighton: Institute of Development Studies.

Sanchez, M.R., et al. 2006. 'Política Externa Como Política Pública: Uma Análise Pela Regulamentação Constitucional Brasileira (1967–1988)', *Revista de Sociologia e Política* 27: 125–143.

Siméant, J. 2015. 'Three Bodies of Moral Economy: The Diffusion of a Concept', *Journal of Global Ethics* 11(2): 163–175.

Six, C. 2009. 'The Rise of Postcolonial States as Donors: A Challenge to the Development Paradigm?', *Third World Quarterly* 30(6): 1103–1121.

Stone, D., O. Porto de Oliveira and L.A. Pal. 2020. 'Transnational Policy Transfer: The Circulation of Ideas, Power and Development Models', *Policy and Society* 39(1): 1–18.

Swyngedouw, E. 2013. 'UN Water Report 2012: Depoliticizing Water', *Development and Change* 44(3): 823–835.

Thompson, L., and P.T. De Wet. 2018. 'BRICS Civil Society Initiatives: Towards the Inclusion of Affected Communities in Collective Development?' *Third World Thematics: A TWQ Journal* 3(5–6): 745–764.

UNDP China. 2018. *Communicating Development Cooperation to Domestic Audiences: Approaches and Implications for South-South Cooperation Providers*. Beijing: UNDP China.

Van der Westhuizen, J. 2012. 'Falling on Fertile Ground? The Story of Emerging Powers' Claims for Redistribution and the Global Poverty Debate', *Global Society* 26: 331–350.

Van der Westhuizen, J., and C.R.S. Milani. 2019. 'Development Cooperation, the International–Domestic Nexus and the Graduation Dilemma: Comparing South Africa and Brazil', *Cambridge Review of International Affairs* 32(1): 22–42.

Von Lieres, B., and L. Piper. 2014. *Mediated Citizenship: The Informal Politics of Speaking for Citizens in the Global South*. London: Palgrave Macmillan.

Waisbich, L.T. 2017. *A Sociedade Civil e o Novo Banco de Desenvolvimento: Aonde Estamos e Para Onde Vamos?* São Paulo: Articulação Sul, Oxfam Brasil.

Waisbich, L.T., and C. Borges. 2019. 'The BRICS' New Development Bank at the Crossroads: Challenges for Building Development Cooperation in the 21st Century', in J. Pupim de Oliveira and J. Yijia (eds.), *International Development Assistance and the BRICS*. London: Palgrave Macmillan, pp. 149–187.

Index

Note: Page numbers followed by "n" refer to notes.

www.ingramcontent.com/pod-product-compliance
Lightning Source LLC
LaVergne TN
LVHW021127300926
849160LV00010B/896
9780367639051